Nadia Nedjah, Ajith Abraham and Luiza de Macedo Mourelle (Eds.)

Computational Intelligence in Information Assurance and Security

Studies in Computational Intelligence, Volume 57

Editor-in-chief
Prof. Janusz Kacprzyk
Systems Research Institute
Polish Academy of Sciences
ul. Newelska 6
01-447 Warsaw
Poland
E-mail: kacprzyk@ibspan.waw.pl

Further volumes of this series
can be found on our homepage:
springer.com

Vol. 33. Martin Pelikan, Kumara Sastry, Erick
Cantú-Paz (Eds.)
*Scalable Optimization via Probabilistic
Modeling*, 2006
ISBN 978-3-540-34953-2

Vol. 34. Ajith Abraham, Crina Grosan, Vitorino
Ramos (Eds.)
Swarm Intelligence in Data Mining, 2006
ISBN 978-3-540-34955-6

Vol. 35. Ke Chen, Lipo Wang (Eds.)
Trends in Neural Computation, 2007
ISBN 978-3-540-36121-3

Vol. 36. Ildar Batyrshin, Janusz Kacprzyk, Leonid
Sheremetor, Lotfi A. Zadeh (Eds.)
*Preception-based Data Mining and Decision Making
in Economics and Finance*, 2006
ISBN 978-3-540-36244-9

Vol. 37. Jie Lu, Da Ruan, Guangquan Zhang (Eds.)
E-Service Intelligence, 2007
ISBN 978-3-540-37015-4

Vol. 38. Art Lew, Holger Mauch
Dynamic Programming, 2007
ISBN 978-3-540-37013-0

Vol. 39. Gregory Levitin (Ed.)
Computational Intelligence in Reliability Engineering,
2007
ISBN 978-3-540-37367-4

Vol. 40. Gregory Levitin (Ed.)
Computational Intelligence in Reliability Engineering,
2007
ISBN 978-3-540-37371-1

Vol. 41. Mukesh Khare, S.M. Shiva Nagendra (Eds.)
*Artificial Neural Networks in Vehicular Pollution
Modelling*, 2007
ISBN 978-3-540-37417-6

Vol. 42. Bernd J. Krämer, Wolfgang A. Halang (Eds.)
Contributions to Ubiquitous Computing, 2007
ISBN 978-3-540-44909-6

Vol. 43. Fabrice Guillet, Howard J. Hamilton (Eds.)
Quality Measures in Data Mining, 2007
ISBN 978-3-540-44911-9

Vol. 44. Nadia Nedjah, Luiza de Macedo
Mourelle, Mario Neto Borges,
Nival Nunes de Almeida (Eds.)
Intelligent Educational Machines, 2007
ISBN 978-3-540-44920-1

Vol. 45. Vladimir G. Ivancevic, Tijana T. Ivancevic
*Neuro-Fuzzy Associative Machinery for Comprehensive
Brain and Cognition Modeling*, 2007
ISBN 978-3-540-47463-0

Vol. 46. Valentina Zharkova, Lakhmi C. Jain
*Artificial Intelligence in Recognition and Classification
of Astrophysical and Medical Images*, 2007
ISBN 978-3-540-47511-8

Vol. 47. S. Sumathi, S. Esakkirajan
*Fundamentals of Relational Database Management
Systems*, 2007
ISBN 978-3-540-48397-7

Vol. 48. H. Yoshida (Ed.)
*Advanced Computational Intelligence Paradigms
in Healthcare*, 2007
ISBN 978-3-540-47523-1

Vol. 49. Keshav P. Dahal, Kay Chen Tan, Peter I. Cowling
(Eds.)
Evolutionary Scheduling, 2007
ISBN 978-3-540-48582-7

Vol. 50. Nadia Nedjah, Leandro dos Santos Coelho,
Luiza de Macedo Mourelle (Eds.)
Mobile Robots: The Evolutionary Approach, 2007
ISBN 978-3-540-49719-6

Vol. 51. Shengxiang Yang, Yew Soon Ong, Yaochu Jin
Honda (Eds.)
*Evolutionary Computation in Dynamic and Uncertain
Environment*, 2007
ISBN 978-3-540-49772-1

Vol. 52. Abraham Kandel, Horst Bunke, Mark Last (Eds.)
*Applied Graph Theory in Computer Vision and Pattern
Recognition*, 2007
ISBN 978-3-540-68019-2

Vol. 53. Huajin Tang, Kay Chen Tan, Zhang Yi
*Neural Networks: Computational Models
and Applications*, 2007
ISBN 978-3-540-69225-6

Vol. 54. Fernando G. Lobo, Cláudio F. Lima
and Zbigniew Michalewicz (Eds.)
Parameter Setting in Evolutionary Algorithms, 2007
ISBN 978-3-540-69431-1

Vol. 55. Xianyi Zeng, Yi Li, Da Ruan and Ludovic Koehl
(Eds.)
Computational Textile, 2007
ISBN 978-3-540-70656-4

Vol. 56. Akira Namatame, Satoshi Kurihara and
Hideyuki Nakashima (Eds.)
Emergent Intelligence of Networked Agents
ISBN 978-3-540-71073-8

Vol. 57. Nadia Nedjah, Ajith Abraham and Luiza de
Macedo Mourella (Eds.)
*Computational Intelligence in Information Assurance
and Security*
ISBN 978-3-540-71077-6

Nadia Nedjah
Ajith Abraham
Luiza de Macedo Mourelle
(Eds.)

Computational Intelligence in Information Assurance and Security

With 100 Figures and 28 Tables

 Springer

Nadia Nedjah
Universidade do Estado do Rio de Janeiro
Faculdade de Engenharia
Sala 5022-D
Rua São Francisco Xavier 524
20550-900, MARACANÃ-RJ
Brazil
E-mail: nadia@eng.uerj.br

Luiza de Macedo Mourelle
Universidade do Estado do Rio de Janeiro
Faculdade de Engenharia
Sala 5022-D
Rua São Francisco Xavier 524
20550-900, MARACANÃ-RJ
Brazil
E-mail: ldmm@eng.uerj.br

Ajith Abraham
Centre for Quantifiable Quality of Service
in Communication Systems (Q2S)
Centre of Excellence
Norwegian University of Science
and Technology
O.S. Bragstads plass 2E
N-7491 Trondheim
Norway
E-mail: ajith.abraham@ieee.org

Library of Congress Control Number: 2007922191

ISSN print edition: 1860-949X
ISSN electronic edition: 1860-9503
ISBN-10   3-540-71077-9 Springer Berlin Heidelberg New York
ISBN-13   978-3-540-71077-6 Springer Berlin Heidelberg New York

Springer is a part of Springer Science+Business Media
springer.com
© Springer-Verlag Berlin Heidelberg 2007

The use of general descriptive names, registered names, trademarks, etc. in this publication does not imply, even in the absence of a specific statement, that such names are exempt from the relevant protective laws and regulations and therefore free for general use.

Cover design: deblik, Berlin
Typesetting by the editors using a Springer LaTeX macro package
Printed on acid-free paper   SPIN: 11971634   89/SPi   5 4 3 2 1 0

Preface

The global economic infrastructure is becoming increasingly dependent upon information technology, with computer and communication technology being essential and vital components of Government facilities, power plant systems, medical infrastructures, financial centers and military installations to name a few. Finding effective ways to protect information systems, networks and sensitive data within the critical information infrastructure is challenging even with the most advanced technology and trained professionals.

This volume provides the academic and industrial community with a medium for presenting original research and applications related to information assurance and security using computational intelligence techniques. The included chapters communicate current research on information assurance and security regarding both the theoretical and methodological aspects, as well as various applications in solving real world information security problems using computational intelligence.

In Chapter 1, which is entitled *Cryptography and Cryptanalysis Through Computational Intelligence*, the authors give a brief introduction to cryptography and Computational Intelligence methods, followed by a short survey of the applications of Computational Intelligence to cryptographic problems. Their contribution in this field is presented. Specifically, some cryptographic problems are viewed as discrete optimization tasks and Evolutionary Computation methods are utilized to address them. The authors show that Artificial Neural Networks are effective in approximating some cryptographic functions. They also present some theoretical issues of Ridge Polynomial Networks and cryptography.

In Chapter 2, which is entitled *Multimedia Content Protection Based on Chaotic Neural Networks*, the author gives a brief introduction to chaotic neural network based data encryption, including stream ciphers, block ciphers and hash functions. Then, he analyzes chaotic neural networks' properties that are suitable for data encryption, such as parameter sensitivity, random similarity, diffusion property, confusion property, one-way property, etc. The author also gives some proposals to design chaotic neural network based cipher or hash function, and uses these ciphers to construct media encryption and authentication methods.

In Chapter 3, which is entitled *Evolutionary Regular Substitution Boxes for Secure Cryptography Using Nash equilibrium*, the authors focus on engineering regular Substitution Boxes or S-boxes that present high non-linearity and low auto-correlation properties using evolutionary computation. There are three properties that need to be optimised: regularity, non-linearity and auto-correlation. The authors exploit the Nash equilibrium-based multi-objective evolutionary algorithm to engineer resilient substitution boxes.

In Chapter 4, which is entitled *Industrial Applications Using Wavelet Packets for Gross Error Detection*, the authors address the Gross Error Detection using uni-variate signal-based approaches and propose an algorithm for the peak noise level determination in measured signals. More specifically, they present developed algorithms and results using two uni-variate, signal-based approaches regarding performance, parameterization, commissioning, and on-line applicability. They base their approach on two algorithms: the Median Absolute Deviation (MAD) a nd the wavelet-based one. Many findings are drawn form the comparison.

In Chapter 5, which is entitled *Immune-inspired Algorithm for Anomaly Detection*, the author presents a new theory: the Danger theory and Dendritic cells, and explores the relevance of those to the application domain of security. He introduces an immune based anomaly detection approach from the abstraction of Danger theory. He also presents the derivation of bio-inspired anomaly detection from the DC functionality with Danger theory, and depicts two examples of how the proposed approach can be applied for computer and network security issues with preliminary results.

In Chapter 6, which is entitled *How to Efficiently Process Uncertainty within a Cyberinfrastructure*, the authors propose a simple solution to the problem of estimating uncertainty of the results of applying a black-box algorithm – without sacrificing privacy and confidentiality of the algorithm.

In Chapter 7, which is entitled *Fingerprint Recognition Using a Hierarchical Approach*, the authors introduce a topology-based approach to fingerprint recognition utilizing both global and local fingerprint features. They also describe a new hierarchical approach to fingerprint matching which can be used to accelerate the speed of fingerprint identification in large databases. Furthermore, the authors propose to apply Radial Basis Functions to model fingerprint's elastic deformation, which greatly increases the system's tolerance to distorted images. They claim that experimental results confirm that the proposed hierarchical matching algorithm achieves very good performance with respect to both speed and accuracy.

In Chapter 8, which is entitled *Smart Card Security*, the authors describe the various attacks that can be applied to smart cards, and the subsequent countermeasures required in software to achieve a secure solution. A case study on the various generations of the European mobile telephone networks is given as an example of how the deployment of countermeasures has changed due to the described attacks.

In Chapter 9, which is entitled *Governance of Information Security: New Paradigm of Security Management*, the author provides a structured approach of security governance to corporate executives. He summarises previous studies on the governance and security management to be able to explain the components and requirements of a governance framework for corporate security. The author provides a governance framework for corporate security, which consists of four domains and two relationship categories.

We are very much grateful to the authors of this volume and to the reviewers for their tremendous service by critically reviewing the chapters. The editors would like also to thank Prof. Janusz Kacprzyk, the editor-in-chief of the Studies in Computational Intelligence Book Series and Dr. Thomas Ditzinger, Springer Verlag, Germany for the editorial assistance and excellent cooperative collaboration to produce this important scientific work. We hope that the reader will share our excitement to present this volume on **Computational Intelligence in Information Assurance and Security** and will find it useful.

November 2006

Nadia Nedjah[1], **Ajith Abraham**[2] and **Luiza M. Mourelle**[1] (Eds.)

[1]State University of Rio de Janeiro, Brazil
[2]IITA Professorship Program, Chung-Ang University, Korea

Contents

List of Figures

List of Tables

1

Cryptography and Cryptanalysis Through Computational Intelligence

E.C. Laskari[1,4], G.C. Meletiou[2,4], Y.C. Stamatiou[3,4], and M.N. Vrahatis[1,4]

[1] Computational Intelligence Laboratory, Department of Mathematics, University of Patras, GR–26110 Patras, Greece
elena@math.upatras.gr, vrahatis@math.upatras.gr
[2] A.T.E.I. of Epirus, Arta, Greece, P.O. Box 110, GR–47100 Arta, Greece
gmelet@teiep.gr
[3] University of Ioannina, Department of Mathematics, GR–45110 Ioannina, Greece istamat@uoi.gr
[4] University of Patras Artificial Intelligence Research Center (UPAIRC), University of Patras, GR–26110 Patras, Greece

The past decade has witnessed an increasing interest in the application of Computational Intelligence methods to problems derived from the field of cryptography and cryptanalysis. This phenomenon can be attributed both to the effectiveness of these methods to handle hard problems, and to the major importance of automated techniques in the design and cryptanalysis of cryptosystems.

This chapter begins with a brief introduction to cryptography and Computational Intelligence methods. A short survey of the applications of Computational Intelligence to cryptographic problems follows, and our contribution in this field is presented. Specifically, some cryptographic problems are viewed as discrete optimization tasks and Evolutionary Computation methods are utilized to address them. Furthermore, the effectiveness of Artificial Neural Networks to approximate some cryptographic functions is studied. Finally, theoretical issues of Ridge Polynomial Networks and cryptography are presented.

The experimental results reported suggest that problem formulation and representation are critical determinants of the performance of Computational Intelligence methods in cryptography. Moreover, since strong cryptosystems should not reveal any patterns of the encrypted messages or their inner structure, it appears that Computational Intelligence methods can constitute a first measure of the cryptosystems' security.

E.C. Laskari et al.: *Cryptography and Cryptanalysis Through Computational Intelligence*, Studies in Computational Intelligence (SCI) **57**, 1–49 (2007)
www.springerlink.com   © Springer-Verlag Berlin Heidelberg 2007

1.1 Introduction

A basic task of cryptography is the transformation, or *encryption*, of a given message into another message which appears meaningful only to the intended recipient through the process of *decryption*. The message that undergoes encryption is called the *plaintext* (or cleartext), while the transformed message is called *ciphertext*. *Cryptanalysis* refers to the process of discovering the plaintext from the ciphertext without knowing the decryption key. A cryptographic algorithm, *cipher*, is a mathematical function employed for the encryption and decryption of messages. Ciphers can be divided into two categories, the *symmetric-key* and the *public-key* ciphers. In symmetric-key ciphers the sender and the receiver of the message secretly choose the key that will be used for encryption and decryption. A drawback of this type of cryptosystems is that it requires prior communication of the key between the sender and the receiver, through a secure channel, before any message is sent.

Public-key ciphers are designed in such a way that the key used for encryption is publicly available and differs from the key used in decryption, which is secret. Although these two keys are functionally interrelated, the computation of the secret key from the public key is computationally intractable. Thus, using the public key anyone can send an encrypted message, but only the owner of the secret key can perform the decryption. Next, we briefly present the cryptographic algorithms that are used in the reported cryptanalysis experiments.

1.1.1 Block ciphers

A block cipher is a function which maps n-bit plaintext blocks to n-bit ciphertext blocks, where n is a chosen blocklength. The function is parameterized by a k-bit key K, which takes values from a subset $\mathcal{K}$, the *key space*, of the set of all k-bit vectors. The function must be invertible to allow unique decryption. Block ciphers can be either symmetric-key or public-key [85].

A *Feistel cipher* is an iterated block cipher based on the repetitive computation of simple functions, called *round functions*, on the input data, for a predetermined number of rounds. The resulting function maps an n–bit plaintext P, to a ciphertext C. In a Feistel cipher the currently computed (by the round function) n–bit word is divided into $(n/2)$–bit parts, the left part L_i and the right part R_i [32]. Then, the ith round, $1 \leqslant i \leqslant r$, has the following effect:

$$L_i = R_{i-1}, \quad R_i = L_{i-1} \oplus F_i(R_{i-1}, K_i), \tag{1.1}$$

where K_i is the subkey used in the ith round (derived from the cipher key K), and F_i is an arbitrary round function for the ith round. After the last round function has been performed, the two halves are swapped and the outcome is the ciphertext C of the Feistel cipher, i.e. $C = (R_r, L_r)$. The encryption procedure of Feistel ciphers is illustrated in Fig. 1.1.

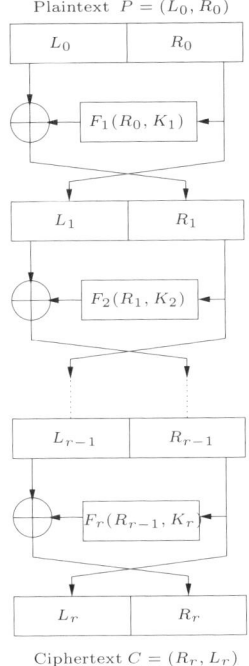

Plaintext $P = (L_0, R_0)$

Ciphertext $C = (R_r, L_r)$

Fig. 1.1. The encryption procedure of Feistel ciphers

On Feistel based cryptosystems the decryption function is simply derived from the encryption function by applying the subkeys, K_i, and the round functions, F_i, in reverse order. This renders the Feistel structure an attractive choice for software and hardware implementations.

One of the most widely used (for non-critical applications) Feistel ciphers is the Data Encryption Standard (DES) [94]. DES is a symmetric-key cryptosystem, meaning that the parties exchanging information possess the same key. It processes plaintext blocks of $n = 64$ bits, producing 64–bit ciphertext blocks. The effective key size is $k = 64$ bits, 8 of which can be used as parity bits. The main part of the round function is the F function, which works on the right half of the data, using a subkey of 48 bits and eight S-boxes. *S-boxes* are mappings that transform 6 bits into 4 bits in a nonlinear manner and constitute the only nonlinear component of DES. The 32 output bits of the F function are XORed with the left half of the data and the two halves are subsequently exchanged. A detailed description of the DES algorithm can be found in [85, 125].

Two of the most powerful cryptanalytic attacks for Feistel based ciphers, rely on the exploitation of specific weaknesses of the S-boxes of the target cryptoalgorithm. These attacks are the *Linear Cryptanalysis* [78, 79] and the *Differential Cryptanalysis* [6, 7], which were successfully applied first to the cryptanalysis of DES. Differential Cryptanalysis (DC) is a *chosen plaintext* attack. In chosen plaintext attacks the opponent has temporary access to the encryption function and thus he/she can choose some plaintexts and construct the corresponding ciphertexts. DC analyzes the effect of particular differences in plaintext pairs on the differences of the resulting ciphertext pairs. These differences can be used to assign probabilities to the possible keys and to identify bits of the key that was used in the encryption process. This method usually works on a number of pairs of plaintexts having a specific difference and relies on the resulting ciphertext pairs only. For cryptosystems similar to DES, the difference is chosen to be a fixed XORed value of the two plaintexts.

To locate the most probable key, DC employs *characteristics*. Note that any pair of encrypted plaintexts is associated with the XOR value of its two plaintexts, the XOR value of its ciphertexts, the XOR values of the inputs of each round in the two encryption executions and the XOR values of the outputs of each round in the two encryption executions. These XOR values form an r-round characteristic [6]. More formally, an r-*round characteristic* is a tuple $\Omega = (\Omega_P, \Omega_\Lambda, \Omega_C)$, where Ω_P and Ω_C are n bit numbers and Ω_Λ is a list of r elements $\Omega_\Lambda = (\Lambda_1, \Lambda_2, \ldots, \Lambda_r)$, each of which is a pair of the form $\Lambda_i = (\lambda_I^i, \lambda_O^i)$, where λ_I^i and λ_O^i are $n/2$ bit numbers and n is the block size of the cryptosystem [6]. A characteristic satisfies the following requirements:

(a) λ_I^1 is the right half of Ω_P,
(b) λ_I^2 is the left half of $\Omega_P \oplus \lambda_O^1$,
(c) λ_I^r is the right half of Ω_C,
(d) λ_I^{r-1} is the left half of $\Omega_C \oplus \lambda_O^r$, and
(e)   for every i, $2 \leqslant i \leqslant r-1$, it holds that $\lambda_O^i = \lambda_I^{i-1} \oplus \lambda_I^{i+1}$.

To each characteristic is assigned the probability of a random pair with the chosen plaintext XOR, Ω_P, having the round XOR values, Λ_i, and the ciphertext XOR, Ω_C, specified in the characteristic. Each characteristic allows the search for a particular set of bits in the subkey of the last round: the bits that enter particular S-boxes, depending on the chosen characteristic. The characteristics that are most useful are those that have a maximal probability and a maximal number of subkey bits whose occurrences can be counted.

DC is a statistical method that rarely fails. A more extended analysis on DC and its results on DES for different numbers of rounds is provided in [6]. DC was the first theoretical cryptanalysis for DES requiring (on average) less steps than the brute force attack, i.e., testing all $2^{56} = 72\,057\,594\,037\,927\,936$ possible keys. Although this number appears prohibitive, a brute-force attack on 56-bit DES, using technology standards of previous decades, has been successfully launched. A specially designed hardware with appropriate software, designed and built by the Cryptography Research, the Advanced Wireless Technologies, and the EFF (Electronic Frontier Foundation) reached a rate of key searches of about 90 billion keys per second. Their prototype, called Deep Crack, contains 29 boards each containing 64 specially designed chips. The achieved key search led to the determination of the key in the RSA DES challenge ($10.000 worth) after an, approximately, 56 hours search effort on July 15 in 1998. Moreover, total cost remained at a relatively low level, below $250.000 (much more lower today) which renders their achievement even more important and worrying as far as the security of 56-bit DES is concerned. However, as if anticipating this attack, the National Institute of Standards and Technology (NIST) had already initiated in 1997 an international contest, accepting proposals for what will become the new standard to replace DES. The contest winner was called *Advanced Encryption Standard* (AES) and it is expected to withstand attacks for a period of at least 30 years, as Miles Smid, the manager of the security technology division of NIST, stated. AES became the government standard and it is also used by private companies (on a royalty-free basis). In 1998, NIST announced the acceptance of fifteen candidate algorithms (in the first round of the process) and resorted to the cryptography community to investigate their security and efficiency. After reviewing the studies and the relevant reports, NIST selected five finalists (Round 2). Among these Rijndael, proposed by Daemen and Rijmen was selected as the new DES. The other four finalists were: MARS (proposed by IBM), RC6 (proposed by RSA Laboratories), Serpent by Anderson, Biham, and Knudsen and Twofish by Schneier, Kelsey, Whiting, Wagner, Hall, and Ferguson. In a third round, NIST concluded that the *Rijndael* cipher should become the Advanced Encryption Standard. Since then, various attacks have been proposed on this cipher but none with devastating effects.

In Sect. 1.4.2 the problem of finding some missing bits of the key that is used in a simple Feistel cipher, namely the Data Encryption Standard with four and six rounds, respectively is studied.

1.1.2 Public key cryptographic schemes

Public key cryptography is intimately related to a number of hard and complex mathematical problems from the fields of computational algebra, number theory, probability theory, mathematical logic, Diophantine's complexity and algebraic geometry. Such problems are the factorization [112], the discrete logarithm [1, 96, 104] and others [86]. Cryptosystems rely on the assumption that these problems are computationally intractable, in the sense that their computation cannot be completed in polynomial time.

Discrete Logarithm Problem (DLP): DLP amounts to the development of an efficient algorithm for the computation of an integer x that satisfies the relation $\alpha^x = \beta$, where α is a fixed primitive element of a finite field $\mathbb{F}_q$ (i.e., α is a generator of the multiplicative group $\mathbb{F}_q^*$ of $\mathbb{F}_q$) and β is a non-zero element of the field. We assume that x is the smallest nonnegative integer with $\alpha^x = \beta$. Then, x is called the *index*, or the *discrete logarithm*, of β. In the special case of a finite field $\mathbb{Z}_p$ of prime order p, a primitive root g modulo p is selected. If u is the smallest nonnegative integer with

$$g^u \equiv h \pmod{p}, \tag{1.2}$$

then u is called the *index*, or the *discrete logarithm*, of h [1, 96, 104].

The security of various public and symmetric key cryptosystems [1, 22, 29, 82, 83, 93, 95, 96, 104, 133], namely the *Diffie–Hellman exchange protocol* [25], the *El Gamal public key cryptosystem*, as well as, *the El Gamal digital signature scheme* [29], relies on the assumption that DLP is computationally intractable.

Diffie–Hellman key Problem (DHP): DHP is defined as follows [22, 80, 133]. Let α be a fixed primitive element of a finite field $\mathbb{F}_q$; x, y, satisfying, $0 \leqslant x, y \leqslant q - 2$, denote the private keys of two users; and $\beta = \alpha^x$, $\gamma = \alpha^y$ represent the corresponding public keys. Then, the problem amounts to computing α^{xy} from β and γ, where α^{xy} is the symmetric key for secret communication between the two users. Consider the special case of the DHP, where $\beta = \gamma$. The term *Diffie–Hellman Mapping* refers to the mapping,

$$\beta = \alpha^x \longmapsto \alpha^{x^2}. \tag{1.3}$$

Diffie–Hellman Mapping Problem (DHMP): The definition of the DHMP follows naturally from the aforementioned definition of DHP. The two problems, DHP and DHMP, are computationally equivalent, as the following relation holds $\alpha^{x^2} \alpha^{y^2} \alpha^{2xy} = \alpha^{(x+y)^2}$, and the computation of α^{xy} from α^{2xy} is feasible (square roots over finite fields).

For the discrete logarithm problem and the Diffie–Hellman key problem, the following theorem holds:

Theorem 1. *Let G be a cyclic group of order m and $G = \langle \alpha \rangle$, where $\alpha^m = e$, and e is the neutral element of the group. Then the well–known cryptosystems of DLP and DHP based on the group G can be represented by matrices of the form*

$$\mathbf{x}^\top W \mathbf{y}, \tag{1.4}$$

where $\mathbf{x}, \mathbf{y}$ are vectors and $W = \{w_{ij}\}_{i,j=1}^m = \alpha^{-ij}$.

Proof. The proof follows by taking into consideration that there exists a prime p such that $m | p - 1$ (or $m | p^n - 1 = q - 1$). Then G can be considered as a subgroup of the multiplicative group of $\mathbb{Z}_p^*$ (or $\mathrm{GF}^*(p, n)$) and according to [22, 68, 81, 93], such a representation exists.

Factorization problem: The factorization problem on the other hand, is related to the RSA cryptosystem and its variants [112]. The security of this cryptosystem relies on the computational intractability of the factorization of a positive integer $N = p \times q$, where p and q are distinct odd primes [85]. The factorization of N is equivalent to determining $\phi(N)$ from N, where $\phi(N) = (p - 1) \times (q - 1)$ [112]. Numerous techniques, including algebraic, number theoretic, soft computing and interpolation methods, have been proposed to tackle the aforementioned problems [1, 22, 62, 96, 120].

In Sects. 1.4.1 and 1.4.3, the DLP, the DHP, the DHMP and the factorization problem are studied in different settings utilizing Evolutionary Computation methods and Artificial Neural Networks, respectively.

1.1.3 Elliptic Curve based cryptosystems

Cryptographic systems based on elliptic curves were proposed in [57, 90] as an alternative to conventional public key cryptosystems. Their main advantage is the use smaller parameters (in terms of bits) compared to the conventional cryptosystems (e.g. RSA). This is due to the apparently increased difficulty of the *Elliptic Curve Discrete Logarithm Problem* (ECDLP), which constitutes the underlying mathematical problem. ECDLP is believed to require more time to solve than the time required for the solution of its finite field analogue, the Discrete Logarithm Problem (DLP). The security of cryptosystems that rely on discrete logarithms, relies on the hypothesis these problems cannot be solved in polynomial time. Numerous techniques that exploit algebraic and number theoretic methods, software oriented methods, as well as, approximation and interpolation techniques [22, 67, 80, 83, 134], have been proposed to speed up the solution of these two types of the discrete logarithm problem.

An *Elliptic Curve* over a prime finite field $\mathbb{F}_p$, where $p > 3$ and prime, is denoted by $E(\mathbb{F}_p)$ and is defined as the set of all pairs $(x, y) \in \mathbb{F}_p$ (points in affine coordinates) that satisfy the equation $y^2 = x^3 + ax + b$, where $a, b \in \mathbb{F}_p$, with the restriction $4a^3 + 27b^2 \neq 0$. These points, together with a special point denoted by $\mathcal{O}$, called *point at infinity*, and an appropriately defined point addition operation form an Abelian group. This is the *Elliptic Curve group* and the point $\mathcal{O}$ is its identity element (see [8, 121] for more details on this group). The *order m of an elliptic curve* is defined as the number of points in $E(\mathbb{F}_p)$. According to Hasse's theorem (see e.g., [8, 121]) it holds that $p + 1 - 2\sqrt{p} \leqslant m \leqslant p + 1 + 2\sqrt{p}$. The *order of a point* $P \in E(\mathbb{F}_p)$ is the smallest positive integer, n, for which $nP = \mathcal{O}$. From Lagrange's theorem, it holds that the order of a point is a divisor of the order of the elliptic curve.

DLP can be described as follows. Let G be any group and h one of its elements. Then, the DLP for G to the base $g \in G$ consists of determining an integer, u, such that $g^u = h$, when the group operation is written as multiplication, or, $ug = h$ when the group operation is written as addition. In groups formed by elliptic curve points the group operation is addition. Therefore, let E be an elliptic curve over a finite field $\mathbb{F}_q$, P a point on $E(\mathbb{F}_q)$ of order n, and Q a point on $E(\mathbb{F}_q)$, such that $Q = tP$, with $0 \leqslant t \leqslant (n-1)$. The ECDLP amounts to determining the value of t. The best algorithms to solve the ECDLP require an exponential number of expected steps, in contrast to the best algorithms known today for the DLP defined over the multiplicative group of $\mathbb{F}_q$, which require sub-exponential time in the size of the used group. In Sect. 1.4.4 the problem of computing the least significant bit of the ECDLP using Artificial Neural Netwroks is studied with interesting results.

1.2 Computational Intelligence Background and Methods

Alan Turing is considered to be the first who conceived the idea of Artificial and Computational Intelligence as early as 1950, when he hypothesized that computers that mimic the processes of the human brain can be developed. This hypothesis implies that any reasoning can be carried out on a large enough deterministic computer. Turing's hypothesis remains a vision, but it has inspired a great amount of research in the effort to embed intelligence to computers.

Although there is no commonly accepted definition, Computational Intelligence (CI) can be considered as the study of adaptive mechanisms that enable intelligent behavior of a system in complex and changing environments [28,30]. These mechanisms exhibit the ability to learn, or adapt to new situations, such that one or more attributes of reason, such as generalization, discovery, association and abstraction, are perceived to be possessed by the system. To enable intelligent behavior, CI systems are often designed to model aspects of biological and natural intelligence. Thus, CI systems are usually hybrids of paradigms such as Evolutionary Computation systems, Artificial Neural Networks and Fuzzy systems, supplemented with elements of reasoning.

1.2.1 Evolutionary Computation

Evolutionary Computation (EC) is a branch of CI that draws its inspiration from evolutionary mechanisms such as natural selection and adaptive behavior, to design optimization and classification methods. Natural selection refers to the survival of the fittest through reproduction. An offspring must retain those characteristics of its parents that are best suited to survive in a given environment. Offsprings that are weak lose the battle of survival. The EC paradigms that form this class are *Genetic Algorithms* (GA), *Genetic Programming* (GP), *Evolutionary Programming* (EP), *Evolution Strategies* (ES) and *Differential Evolution* (DE). The social and adaptive behavior of animals organized in groups inspired the development of another class of EC methods, namely *Swarm Intelligence* (SI). These methods model the social procedures of living organisms that are organized into groups and act for a common cause. Typical examples are the search for food mechanisms of bird flocks, fish schools and ant colonies. The study of many biological processes of social and adaptive behavior led to the opinion that social sharing of information among the individuals of a population can generate an evolutionary advantage [28]. Paradigms of EC that belong to this class of methods are the *Particle Swarm Optimization* (PSO) method and the *Ant Colony Optimization* (ACO) method. In the following sections a brief description of each paradigm of EC is given. Since, the PSO and DE methods will be used in our experiments in Sect. 1.4, they are more thoroughly described.

Genetic Algorithms

The experimentation of biologists in simulating natural genetic systems using computers gave rise to Genetic Algorithms (GA). John Holland is regarded as the creator of the field of GAs. He studied machine intelligence and machine learning and developed the abilities of GAs to artificial systems [45]. These systems had the ability to adapt to changes of the environment and also exhibited self–adaptation in the sense that they could adjust their operations according to their interaction with the environment. Among the innovations of Holland was the use of a population of individuals for the search procedure instead of a single search point.

The basic concepts of GAs is natural evolution and genetic inheritance. In natural evolution each biological specie has to search for the most appropriate adaptations to a complex and changing environment to ensure its survival. GAs are based on the idea that the knowledge and experience that a specie gains passes in the chromosomes of its members. For this reason the vocabulary used for GAs is that of genetics. Thus, the individuals of the population are called *chromosomes* or *genotypes*. Each chromosome consists of parts called genes and each of them is responsible for the inheritance of one or more characteristics. The evolution procedure of a population of chromosomes corresponds to a search on the space of possible problem solutions and has to balance between two different scopes, the *exploitation* of the best solutions and the *exploration* of the search space. The evolution procedure of GAs is implemented using two operators, *crossover* and *mutation*. These operators alter chromosomes to produce better ones. The selection of the new population is completed using as a criterion a fitness measure. Regarding the representation of the chromosomes, GAs usually employ binary representation, but GA methods that use other arithmetic systems, including floating point numbers, have also been developed. GAs have been successfully applied to optimization problems arising in different fields such as applied mechanics and design, time–scheduling, the traveling salesman's problem, optimal control and robotics, and economics among others [3, 24, 34, 38, 87].

Evolutionary Programming

The field of Evolutionary Programming (EP) was developed by Larry Fogel [35] parallel to that of GAs. The aim of EP was the evolution of Artificial Intelligence by predicting the changes of the environment. The environment in EP is described as a sequence of symbols from a finite set and the evolution algorithm provides as output a new symbol. This symbol has to maximize the fitness function that is used as a measure for the accuracy of the prediction. For the representation of each individual of the population finite state machines were chosen. Evolutionary Programming, just like GAs, uses the principle of the selection of the fittest for the new population, but only the mutation operator is used for altering the individuals of the population. To this initial version of EP two more basic concepts have been added. The first regards the ability of handling continuous parameters in addition to the discrete ones, and the second is the ability of self–adaptation. Using these new advances, EP can address optimization and classification problems with applications in several scientific fields as for example economics [33, 34].

Evolution Strategies

In the 1970s, Ingo Rechenberg and Hans-Paul Schwefel used the idea of mutation trying to obtain the optimal design for a sequence of joints in a liquid transition pipe. The classical optimization techniques that make use of the gradient of the fitness function were unable to handle the problem and the only solution was the experimentation with mutation. Using mutation they caused a small perturbation to the best existing problem solutions in order to explore in a stochastic manner the neighborhoods in the search space of the problem. This experimentation was the beginning of the development of Evolution Strategies, which were established in 1973 [109]. Evolution Strategies can be considered as evolutionary programs that use floating point representation and employ a *recombination* and a mutation operator. They have been used for the solution of several optimization problems with continuously changing parameters and they have been recently extended for discrete problems [42].

Genetic Programming

Genetic Programming (GP) was developed more recently by Koza [61]. The idea behind GP is the following: instead of constructing an evolutionary program to solve the problem, to locate in the space of computational programs the most proper one for the specific case. GP provides means to achieve this goal. A population of executable computational programs is created and every individual program competes with the rest. The non efficient programs become idle while the best ones reproduce by means of operators such as crossover and mutation. The evaluation of the programs is done using a fitness on a predefined set of problems.

Differential Evolution

The Differential Evolution algorithm (DE) [126] is a parallel direct numerical search method, that utilizes N, D–dimensional parameter vectors $x_{i,G}$, $i = 1, 2, \ldots, N$, as a population for each iteration (generation) of the algorithm. At each generation, the *mutation* and *crossover* (*recombination* [103, 127]) operators are applied on the individuals, to produce a new population, which is subsequently subjected to the selection phase.

For each vector $x_{i,G}$, $i = 1, 2, \ldots, N$, a *mutant vector* is generated through the following equation:

$$v_{i,G+1} = x_{r_1,G} + F\left(x_{r_2,G} - x_{r_3,G}\right), \tag{1.5}$$

where $r_1, r_2, r_3 \in \{1, 2, \ldots, N\}$, are random indexes, mutually different and different from i, and $F \in (0, 2]$. Consequently, N must be greater than, or equal to, 4. Following the mutation phase, the crossover operator is applied on the mutant vector yielding the *trial vector*, $u_{i,G+1} = (u_{1i,G+1}, u_{2i,G+1}, \ldots, u_{Di,G+1})$, where,

$$u_{ji,G+1} = \begin{cases} v_{ji,G+1}, & \text{if } (\text{randb}(j) \leqslant CR) \text{ or } j = \text{rnbr}(i), \\ x_{ji,G}, & \text{if } (\text{randb}(j) > CR) \text{ and } j \neq \text{rnbr}(i), \end{cases} \tag{1.6}$$

for $j = 1, 2, \ldots, D$; where $\text{randb}(j)$, is the jth evaluation of a uniform random number generator in the range $[0, 1]$; CR is the (user specified) crossover constant in the range $[0, 1]$; and $\text{rnbr}(i)$ is a randomly chosen index from the set $\{1, 2, \ldots, D\}$. To decide whether or not the vector $u_{i,G+1}$ will be a member of the population of the next generation, it is compared to the initial vector $x_{i,G}$. Thus,

$$x_{i,G+1} = \begin{cases} u_{i,G+1}, & \text{if } f(u_{i,G+1}) < f(x_{i,G}), \\ x_{i,G}, & \text{otherwise.} \end{cases} \tag{1.7}$$

The DE algorithm that utilizes the mutation operator of (1.5) is called the standard variant of the DE algorithm. Different mutation operators define the other variants of the DE algorithm. The mutation operators that have been applied with promising results [126], are the following:

$$v_{i,G+1} = x_{\text{best},G} + F(x_{r_1,G} - x_{r_2,G}), \tag{1.8}$$

$$v_{i,G+1} = x_{i,G} + F(x_{\text{best},G} - x_{i,G}) + F(x_{r_1,G} - x_{r_2,G}), \tag{1.9}$$

$$v_{i,G+1} = x_{\text{best},G} + F(x_{r_1,G} + x_{r_2,G} - x_{r_3,G} - x_{r_4,G}), \tag{1.10}$$

$$v_{i,G+1} = x_{r_1,G} + F(x_{r_2,G} + x_{r_3,G} - x_{r_4,G} - x_{r_5,G}), \tag{1.11}$$

where, $x_{\text{best},G}$, corresponds to the best individual of the Gth generation, r_1, r_2, r_3, r_4, $r_5 \in \{1, 2, \ldots, N\}$, are mutually different random indexes and $x_{i,G}$ is the current individual of generation G.

Particle Swarm Optimization

Particle Swarm Optimization (PSO) method is a population–based algorithm that exploits a population of individuals, to identify promising regions of the search space. In this context, the population is called *swarm* and the individuals are called *particles*. Each particle moves with an adaptable velocity within the search space, and retains in its memory the best position it ever encountered. In the *global* variant of the PSO the best position ever attained by all individuals of the swarm is communicated to all the particles. In the *local* variant, each particle is assigned to a neighborhood consisting of a prespecified number of particles. In this case, the best position ever attained by the particles that comprise the neighborhood is communicated among them [28].

Assume a D–dimensional search space, $\mathbf{S} \subset \mathbb{R}^D$, and a swarm of N particles. The ith particle is in effect a D–dimensional vector $X_i = (x_{i1}, x_{i2}, \ldots, x_{iD})^\top$. The *velocity* of this particle is also a D–dimensional vector, $V_i = (v_{i1}, v_{i2}, \ldots, v_{iD})^\top$. The *best previous position* ever encountered by the i–th particle is a point in $\mathbf{S}$, denoted by $P_i = (p_{i1}, p_{i2}, \ldots, p_{iD})^\top$. Assume g, to be the index of the particle that attained the best previous position among all the individuals of the swarm (global variant of PSO) or among all individuals of the neighborhood of the i-th particle (local variant of PSO).

Then, according to the *constriction factor* version of PSO the swarm is manipulated using the following equations [21]:

$$V_i^{(t+1)} = \chi \left(V_i^{(t)} + c_1 r_1 \left(P_i^{(t)} - X_i^{(t)} \right) + c_2 r_2 \left(P_g^{(t)} - X_i^{(t)} \right) \right), \quad (1.12)$$

$$X_i^{(t+1)} = X_i^{(t)} + V_i^{(t+1)}, \qquad (1.13)$$

where $i = 1, 2, \ldots, N$; χ is the constriction factor; c_1 and c_2 denote the *cognitive* and *social* parameters respectively; r_1, r_2 are random numbers uniformly distributed in the range $[0, 1]$; and t, stands for the counter of iterations. The value of the constriction factor is typically obtained through the formula $\chi = 2\kappa / |2 - \varphi - \sqrt{\varphi^2 - 4\varphi}|$, for $\varphi > 4$, where $\varphi = c_1 + c_2$, and $\kappa = 1$. The default parameter values found in the literature [21] are $\chi = 0.729$ and $c_1 = c_2 = 2.05$. Different configurations of χ as well as a theoretical analysis of the derivation of the above formula can be found in [21].

In a different version of PSO a parameter called *inertia weight* is used, and the swarm is manipulated according to the formulae [28, 52, 118]:

$$V_i^{(t+1)} = w V_i^{(t)} + c_1 r_1 \left(P_i^{(t)} - X_i^{(t)} \right) + c_2 r_2 \left(P_g^{(t)} - X_i^{(t)} \right), \qquad (1.14)$$

$$X_i^{(t+1)} = X_i^{(t)} + V_i^{(t+1)}, \qquad (1.15)$$

where $i = 1, 2, \ldots, N$; and w is the inertia weight, while all other variables are the same as in the constriction factor version. There is no explicit formula for the determination of the factor w, which controls the impact of the previous history of velocities on the current one. However, since a large inertia weight facilitates global exploration (searching new areas), while a small one tends to facilitate local exploration (fine–tuning the current search area), it appears intuitively appealing to initially set it to a large value and gradually decrease it to obtain more refined solutions. The superiority of this approach against the selection of a constant inertia weight, has been experimentally verified [118]. An initial value around 1.2 and a gradual decline toward 0.1 is considered a good choice for w. Proper fine–tuning of the parameters c_1 and c_2, results in faster convergence and alleviation of local minima. As default values, $c_1 = c_2 = 2$ have been proposed, but experimental results indicate that alternative configurations, depending on the problem at hand, can produce superior performance [52, 98].

In order to avoid velocities from assuming large values that lead to fluctuation of the particles over the search space, and destroy the dynamic of the method, a maximum value, V_{max}, is set for each coordinate of the velocity.

Typically, the swarm and the velocities, are initialized randomly in the search space. For more sophisticated techniques, see [97,99]. The performance of the PSO method for the Integer Programming problem and the Minimax problem was studied in [71,72], respectively, with very promising results.

Ant Colony Optimization

The Ant Colony Optimization (ACO) algorithm is a Swarm Intelligence method for tackling, in general, Combinatorial Optimization problems, like the traveling salesman problem and telecommunications scheduling. It exploits a population of members called *artificial ants* and it has been inspired from experiments with real ant colonies. In these experiments it was discovered that after a small time interval groups of ants choose the shortest between two routes to transfer food to their nest. This ability becomes possible by a chemical substance, called *pheromone*, which ants leave in the environment, that serves as an indirect communication mechanism. Thus, at the beginning the route chosen by ants appears to be random but with the progress of time the possibility of choosing the shortest path becomes higher as the quantity of pheromone on this path increases faster compared to the quantity of pheromone on longer paths. This simple idea is implemented by the ACO methods to locate solutions and address hard optimization problems [10,27].

1.2.2 Artificial Neural Networks

The complex and parallel functionality of the human brain has motivated the design of *Artificial Neural Networks* (ANNs). An ANN can be considered as a massively parallel distributed processor, comprised of simple units called *neurons*, and characterized by an inherent ability to acquire knowledge from data through a learning process. Knowledge is stored at the interneuron connection strengths, called *weights*, making it thus available for use [41]. Each artificial neuron implements a local computation. The output of this computation is determined by the neuron's input and its activation function. The overall functionality of a network is determined by its *topology* (architecture), i.e. the number of neurons and their interconnection pattern, the training algorithm applied, and its neuron characteristics [46,102].

ANNs can be categorized based on their topology, their functionality, their training methods, and other characteristics. Regarding their topology, the most simple ANNs have only one layer of neurons and are called single-layer ANNs, while the ones with more than one layers of neurons are called multi-layer ANNs. Furthermore, ANNs with acyclic interneuron connections are called Feedforward Neural Networks (FNNs), while those with feedback loops are called Recurrent Neural Networks (RNNs). The most commonly used

ANNs are FNNs. A Feedforward Neural Network is a network with acyclic and one-way directed interneuron connections, where neurons can be grouped into layers. Thus, the network's topology can be described by a series of integers each representing the number of units that belong to the corresponding layer.

The functionality of ANNs is based on the type of neurons they consist of, and their activation function. In general, the are two types of neurons, summing and product neurons. Summing neurons apply their activation function over the sum of the weighted inputs, while product neurons apply their activation function over the product of the weighted inputs (see Sect. 1.5). The activation function determines the output of the neuron, and several types of activation functions can be used. The most commonly encountered ones are the *linear function* (1.16), the *threshold function* (1.17), the *sigmoid function* (1.18), the *hyperbolic tangent function* (1.19) and the *Gaussian function* (1.20).

$$f_1(x) = \alpha x, \tag{1.16}$$

$$f_2(x) = \begin{cases} \alpha_1, & \text{if } x \geqslant \theta, \\ \alpha_2, & \text{if } x < \theta, \end{cases} \tag{1.17}$$

$$f_3(x) = \frac{1}{1 + e^{-\lambda_1 x}}, \tag{1.18}$$

$$f_4(x) = \tanh(\lambda_2 x), \tag{1.19}$$

$$f_5(x) = e^{-x^2/\sigma^2}, \tag{1.20}$$

where $\alpha, \alpha_1, \alpha_2, \theta, \lambda_1, \lambda_2$ are constants and σ^2 is the variance of the Gaussian distribution. The training methods for ANNs can be divided into three categories, *supervised learning* methods in which case the ANNs must adapt to given data so as to produce a specific output; *unsupervised learning* where ANNs have to discover patterns on the input data; and *reinforcement learning* that aims at rewarding ANNs for good performance and penalize them otherwise [30, 39, 41, 58].

In the case of supervised learning, the goal of training is to assign to the weights (free parameters) of the network, W, values such that the difference between the desired output (target) and the actual output of the network is minimized. The adaptation process starts by presenting to the network a series of patterns for which the desired outputs are a priori known, and computing a total error function $E = \sum_{k=1}^{P} E_k$. In this equation, P is the number of patterns and E_k is the partial network error with respect to the kth pattern. For the computation of the partial network error a variety of error (distance) functions can be used [74, 132]. Usually, it is computed by summing the squared difference between the actual network outputs and the desired outputs for the corresponding pattern. The training patterns can be presented numerous times to the network. Each pass of all the patterns that belong to the training set, T, is called a *training epoch*. The total number of epochs required can be considered as the speed of the training algorithm. Several training algorithms can be found in [41, 73, 75, 76, 103, 110, 130].

The computational power of neural networks derives from their parallel and distributed structure and their inherent ability to adapt to specific problems, learn, and generalize. These characteristics allow ANNs to solve complex problems. In [46, 131] the following statement has been proved: *"Standard feedforward networks with only a single hidden layer can approximate any continuous function uniformly on any compact set and any measurable function to any desired degree of accuracy"*. It has also been proved [102] that *"a single hidden layer feedforward network with a fixed number of units in the hidden layer, has a lower bound on the degree of the approximation of any function"*. The lower bound obstacle can be alleviated if more than one hidden layers are used. Mairov and Pincus in [102] have proved that, *"on the unit cube in $\mathbb{R}^n$ any continuous function can be uniformly approximated, to within any error by using a two hidden layer network having $2n + 1$ units in the first layer and $4n + 3$ units in the second"*. Furthermore, Anthony in [2] has proved that *"there is a 2-layer threshold network capable of computing any Boolean function"*. These results imply that any lack of success in applications can be attributed to inadequate training, an insufficient number of hidden units, or the lack of a deterministic relationship between input and target.

ANNs have been applied in several scientific fields and addressed efficiently and effectively a number of hard and complex problems. Some classes of ANN applications are *function approximation*, aiming at learning the functional relationship between the inputs and the desired output, *optimization*, i.e., finding the optimal parameter values in an optimization problem, *data mining*, aiming at discovering hidden patterns in data, *classification*, i.e. prediction of the class of an input vector, *pattern matching*, where the scope is to produce a pattern that is best associated with a given input vector, *pattern completion*, where the scope is to complete missing parts of a given input vector, and *control*, where, given an input vector, an appropriate action is suggested [30].

1.2.3 Fuzzy systems

Traditional set theory and binary-valued logic both require two values of parameters, be part of a set or not, and 0 or 1, respectively. Human reasoning, however, includes a measure of uncertainty, and hence is not exact. Fuzzy sets and fuzzy logic allow what is referred to as approximate reasoning. With *fuzzy sets*, an element belongs to a set with a certain degree of certainty. *Fuzzy logic* allows reasoning with these uncertain facts to infer new facts, with a degree of certainty associated with each fact. In a sense, fuzzy sets and fuzzy logic allow the modeling of common sense [30]. The uncertainty in fuzzy systems is referred to as non statistical uncertainty, which should not be confused with statistical uncertainty. Statistical uncertainty is based on the laws of probability, whereas non statistical uncertainty is based on vagueness, imprecision and/or ambiguity. Statistical uncertainty is resolved through observations. For example, when a coin is tossed we are certain what the outcome is, while before tossing the coin, we know that the probability of each outcome is 50%. Nonstatistical uncertainty, or fuzziness, is an inherent property of a system and cannot be altered or resolved by observation. Fuzzy systems have been applied to control systems, gear transmission and braking systems in vehicles, controlling lifts, home appliances, and controlling traffic signals, among others [30, 128].

1.3 Review of Cryptography and Cryptanalysis Through Computational Intelligence

Computational Intelligence methods have been successfully applied in numerous scientific fields. Evolutionary Computation (EC) algorithms share a common characteristic, namely that they do not require good mathematical properties, such as continuity or differentiability, for the objective function of the underlying problem. Therefore, they are applicable to hard real–world optimization problems that involve discontinuous objective functions and/or disjoint search spaces [34, 52, 117]. Artificial Neural Networks (ANNs) have also been applied to many scientific fields and problem classes and provided very promising results, due to their parallel and distributed structure and their inherent ability to adapt, learn and generalize. The use of automated techniques in the design and cryptanalysis of cryptosystems is desirable as it minimizes the need for time-consuming human interaction with the search process [17]. However, due to its nature the field of cryptography and cryptanalysis is quite demanding and complex. Thus, the application of an efficient and effective tool such as Computational Intelligence (CI) to the field of cryptology comes naturally. A brief survey of the research relating the two fields follows.

The works of Peleg and Rosenfeld [100] in 1979, Hunter and McKenzie [47] in 1983, Carrol and Martin [13] in 1986 and King and Bahler [53] in 1992, that used relaxation algorithms for breaking simple substitution ciphers, can be considered as predecessors to the application of EC methods in cryptanalysis. In 1993 Spillman et al. [123, 124] and Mathews [77] introduced the use of genetic algorithms for addressing simple substitution, transposition and knapsack ciphers, while later in the same year Forsyth and Safavi-Naini [36] proposed the simulated annealing method for attacking a simple substitution algorithm. In 1995, Jakobsen [49] proposed some simplified hill-climbing techniques for addressing the problems of [36, 124] and in 1996 Vertan and Geangala [129] used genetic algorithms for breaking the Merkle–Hellman cryptosystem. Also, in 1997 Bagnall et al. [4] presented a ciphertext-only attack for a simplified version of an Enigma rotor machine using genetic algorithms.

In 1998 A. Clark proposed in his Ph.D. thesis [17] the tabu search algorithm for cryptanalysis and compared several heuristic techniques, including genetic algorithms, for breaking classical cryptosystems. In his thesis, it was also proved that the knapsack cipher attack of [123] was flawed and, furthermore, Millan, A. Clark and Dawson proposed the construction of Boolean functions with good cryptographic properties utilizing smart hill-climbing techniques and genetic algorithms [88,89]. Continuing the work of Millan, A. Clark and Dawson, in [18] J. Clark and Jacob presented a two stage optimization for the design of Boolean functions and more recently they proposed new attacking techniques of cryptographic primitives based on fault injection and timing analysis which are effective in breaking a specific kind of identification schemes using simulated annealing [19]. In [12] Burnett et al. designed the S-boxes of MARS, one of the five AES finalists, using hill-climbing and genetic algorithms. Also, J. Clark et al. in [20] provided an improvement for the design of S-boxes using simulated annealing. In 2002 Hernández et al. proposed a new cryptanalytic technique for TEA with reduced number of rounds, which also proved to be useful in distinguishing a block cipher from a random permutation, by applying genetic algorithms [43,44]. Finally, in 2004 Barbieri et al. [5] described a method for generating good linear block error-correcting codes that uses genetic algorithms, following the idea of genetic approach to code generation of Dontas and Jong [26].

Over the last fifteen years, just a few research studies have appeared to relate ANNs with cryptography and cryptanalysis. In 1991 Rivest wrote a survey article about the relationship between cryptography and machine learning [111], emphasizing on how these fields can contribute ideas to each other. Blum, Furst, Kearns and Lipton in [9] presented how to construct good cryptographic primitives based on problems in learning that are believed to be intractable. Working on the same concept, Pointcheval in [105, 106] used an NP-Complete problem based on ANNs for the design of certain type of secure identification problems but later Knudsen and Meier in [56] demonstrated that this scheme is less secure than what was previously believed. In 1998, Ramzan in his Ph.D. thesis [107] broke the Unix Crypt cryptosystem, a simlified variant of the Enigma cryptosystem, using ANNs. In 2001, an application of cryptology to the field of ANNs was proposed by Chang and Lu [14]. Specifically, they proposed oblivious polynomial evaluation protocols that can operate directly to floating point numbers and gave as example the oblivious learning of an ANN. Also, in [135] a general paradigm for building ANNs for visual cryptography is presented. In [50, 51] Karras and Zorkadis used Feedforward and Hopfield neural networks to improve and strengthen traditional pseudorandom stream generators for the secure management of communication systems. ANNs have also been used for the development of a new key exchange system which is based on a new phenomenon, the synchronization of ANNs [54, 55, 113]. The synchronization of ANNs is a kind of mutual training of ANNs on common inputs. However, in [55] it was shown that this key exchange protocol can be broken in three different ways, using genetic algorithms, genetic attack and probabilistic attack, respectively. Lately, in [91,115,116] some techniques for the improvement of the previously proposed protocol were presented. Finally, the idea of applying ANNs for the construction of S-boxes was presented in [60].

1.4 Applying Computational Intelligence in Cryptanalysis

In the following sections our results obtained from the application of CI methods in the cryptanalysis of known cryptosystems are presented. Specifically, in the first section cryptographic problems derived from classical public key cryptosystems are formulated as discrete optimization tasks and EC methods are applied to address them. In the next section, EC methods are considered for the partial cryptanalysis of Feistel ciphers. The effectiveness of ANNs for classical cryptographic problems and problems of elliptic curve cryptography, follow. Lastly, the relationship between a specific class of ANNs, namely the Ridge Polynomial Networks, and theoretical results of cryptography is presented.

1.4.1 Cryptanalysis as Discrete Optimization Task

In this section three problems encountered in the field of cryptology are formulated as discrete optimization tasks and two EC algorithms, namely PSO method and DE algorithm, are applied for their cryptanalysis. The reported results suggest that the formulation of the problems as discrete optimization tasks preserves their complexity which makes it difficult for the methods to extract pieces of information [64,70]. This fact suggests that the main issue when using EC methods in cryptanalysis is the proper definition of the fitness function, i.e., avoiding the deceptive landscapes that lead in results not better than random search, which was also later mentioned in [48]. Thus, the first conclusion derived by these experiments is that, due to the proven complexity of the cryptographic problems, when EC methods are applied to cryptanalysis special attention must be paid to the design of the fitness function so as to include as much information as possible for the target problem. The second conclusion is that EC methods (and CI methods in general) can be used as a quick practical assessment for the efficiency and the effectiveness of proposed cryptographic systems. Specifically, since strong cryptosystems must not reveal any patterns of the encrypted messages or their inner structure (as this could lead to their cryptanalysis), CI methods can be used as a first measure for the evaluation of new cryptographic schemes before more formal methods (which may be complex to apply) are employed for their analysis.

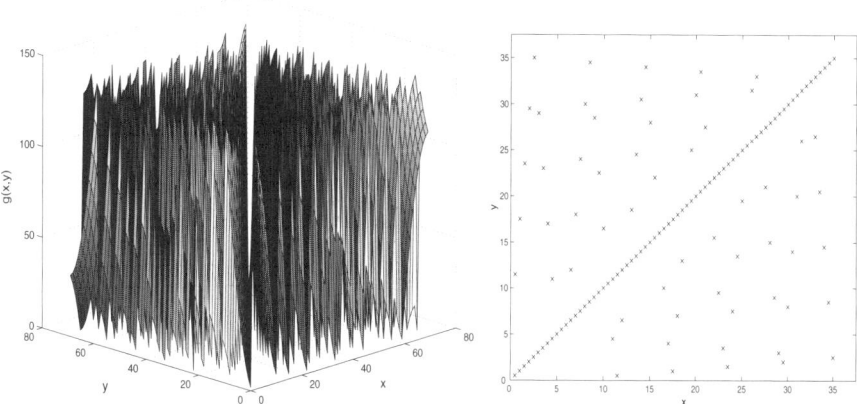

Fig. 1.2. (a) Plot of function $g(x,y) = x^2 - y^2(\text{mod } N)$, for $N = 143$ and (b) contour plot of function $g(x,y) = x^2 - y^2(\text{mod } N)$, for $N = 143$ at value $g = 0$

Problem Formulation

All three problems considered below are derived from the factorization problem described in Sect. 1.1.2. The first problem is defined as follows: given a composite integer N, find pairs of x, $y \in \mathbb{Z}_N^*$, such that $x^2 \equiv y^2(\text{mod } N)$, with $x \not\equiv \pm y(\text{mod } N)$. This problem is equivalent to finding non-trivial factors of N, as N divides $x^2 - y^2 = (x - y)(x + y)$, but N does not divide either $x - y$ or $x + y$. Thus, the $\gcd(x - y, N)$ is a non-trivial factor of N [85].

We formulate the problem as a discrete optimization task by defining the minimization function $f : \{1, 2, \ldots, N{-}1\} \times \{1, 2, \ldots, N{-}1\} \mapsto \{0, 1, \ldots, N{-}1\}$, with

$$f(x,y) = x^2 - y^2 \quad (\text{mod } N), \tag{1.21}$$

subject to the constraints $x \not\equiv \pm y(\text{mod } N)$. The constraint $x \equiv -y(\text{mod } N)$ can be incorporated in the problem by changing the function domain. In this case, the problem reduces to minimizing the function $g : \{2, 3, \ldots, (N - 1)/2\} \times \{2, 3, \ldots, (N - 1)/2\} \mapsto \{0, 1, \ldots, N{-}1\}$, with

$$g(x,y) = x^2 - y^2 \quad (\text{mod } N), \tag{1.22}$$

subject to the constraint $x \not\equiv y(\text{mod } N)$. This is a 2-dimensional minimization problem and the global minimum of the function g is zero. For simplicity, we will call the minimization problem of function g, i.e., finding a global minimizer (x^*, y^*) of the function g, subject to the constraint $x \not\equiv y(\text{mod } N)$, as **Problem 1**. An illustration of the function $g(x,y)$ for $N = 11 \times 13 = 143$ is depicted in Fig. 1.2 (a), and the contour plot of function g at the global minimum $g(x,y) = 0$ is shown in Fig. 1.2 (b).

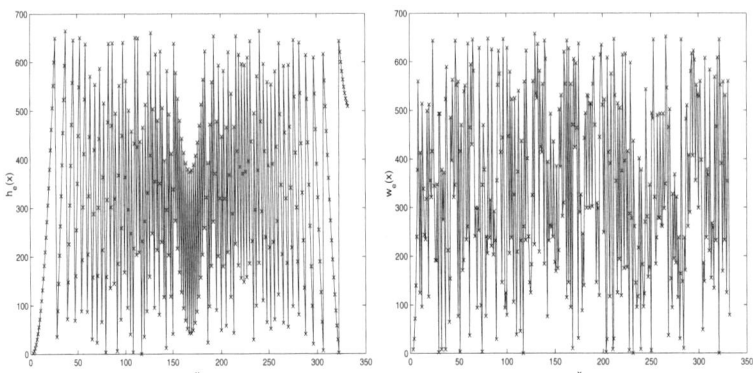

Fig. 1.3. (a) Plot of function $h_e(x) = (x-1)(x-2)(\text{mod } N)$, for $N = 667$ and (b) plot of function $w_e(x) = (x+1)(x-1)(x-2)(\text{mod } N)$, for $N = 667$

We also consider the following minimization problems. Let us define the minimization function $h : \{1, 2, \ldots, N{-}1\} \mapsto \{0, 1, \ldots, N{-}1\}$, with

$$h(x) = (x-a)(x-b) \ (\text{mod } N), \tag{1.23}$$

where a, b non–zero integers and $x \not\equiv a(\text{mod } N)$, $x \not\equiv b(\text{mod } N)$. A test case of this problem is the function

$$h_e(x) = (x-1)(x-2) \ (\text{mod } N), \tag{1.24}$$

where $x \not\equiv 1(\text{mod } N)$ and $x \not\equiv 2(\text{mod } N)$. This is 1-dimensional minimization problem with global minimum zero. We will refer to the minimization of function $h_e(x)$, subject to the constraints $x \not\equiv 1(\text{mod } N)$ and $x \not\equiv 2(\text{mod } N)$, as **Problem 2**. Figure 1.3 (a) depicts the function $h_e(x)$ for the small value of $N = 23 \times 29 = 667$.

In a more general setting, we can consider the minimization of the function

$$w(x) = (x-a)(x-b) \cdots (x-m) \ (\text{mod } N), \tag{1.25}$$

where $x \in \{0, 1, \ldots, N{-}1\}$ and $x \not\equiv \{a, b, \ldots, m\}(\text{mod } N)$. We study the case function

$$w_e(x) = (x+1)(x-1)(x-2) \ (\text{mod } N), \tag{1.26}$$

with $x \not\equiv \{-1, 1, 2\}(\text{mod } N)$. We will refer to the 1-dimensional minimization of function $w_e(x)$, subject to the constraints with $x \not\equiv \{-1, 1, 2\}(\text{mod } N)$, as **Problem 3**. In Fig. 1.3 (b) an illustration of the function $w_e(x)$ for the small value $N = 23 \times 29$ is shown.

Experimental Setup and Results

Both Particle Swarm Optimization (PSO) [21] and Differential Evolution (DE) [126] methods are applied on the problems formulated in the previous section (Problems 1,2,3) and compared with the simple random search technique. The global and local PSO variants of both the inertia weight and the constriction factor versions, as well as the DE variants with the mutation operators of (1.5) and (1.8), are used. The typical parameter values for the PSO variants are used (see Sect. 1.2), and the local variant of the PSO is tested for neighborhood size equal to 1. For the PSO, preliminary experiments on the specific problems indicated that the value of maximum velocity V_{max} of the particles affects its performance significantly. The most promising results were produced using the values $V_{max} = ((N - 7)/10, (N - 7)/10)$ for Problem 1, and the value $V_{max} = (N - 4)/5$ for Problems 2 and 3, and therefore they are adopted in all the experiments. The parameters of the DE algorithm are set at the values $F = 0.5$ and $CR = 0.5$. In all cases, the populations are constrained to lie within the feasible region of the corresponding problem.

For the minimization of function g (Problem 1), the performance of the methods is investigated for several values of N, in the range $N = 199 \times 211 = 41\,989$ up to $N = 691 \times 701 = 484\,391$. For each value of N considered, 100 independent runs are performed. The corresponding results are shown in Table 1.1. In this table, PSOGW corresponds to the global variant of PSO method with inertia weight; PSOGC is the global variant of PSO with constriction factor; PSOLW is PSO's local variant with inertia weight; PSOLC is PSO's local variant with constriction factor, DE1 corresponds to the DE algorithm with the mutation operator of (1.5) and DE2 to the DE algorithm with the mutation operator of (1.8). Random search results are denoted as RS. A run is considered to be successful if the algorithm identifies a global minimizer within a prespecified number of function evaluations. The function evaluations threshold is taken equal to the cardinal of integers in the domain of the target function. The success rates of each algorithm, i.e. the fraction of times it achieved a global minimizer within the prespecified threshold, the minimum number, the median, the mean value and the standard deviation of function evaluations (F.E.) needed for success, are also reported in the table.

The experimental results of Table 1.1 indicate that the variants of the PSO method outperform in success rates the variants of the DE method over these problem instances (i.e., different values of the parameter N) and with the parameter setup used. Moreover, the performance of the DE method decreases as the value of N increases while PSO appears to be more stable with respect to this parameter. However, in contrast to the known behavior of EC methods, the best success rates achieved are relatively low (around 50%) and the random search technique (RS) outperforms both EC methods and their variants. This fact suggests that the almost random behavior of the specific kind of problem makes it quite difficult for the methods to extract knowledge about their dynamics. In the cases where EC methods located a global minimizer they

Table 1.1. Results for the minimization of function g (see (1.22))

N	Method	Suc.Rate	mean F.E.	St.D. F.E.	median F.E.	min F.E.
	PSOGW	56%	8844.643	5992.515	8325.000	660
	PSOGC	48%	7149.375	5272.590	5355.000	330
	PSOLW	51%	8329.412	6223.142	7050.000	270
$N = 199 \times 211$	PSOLC	51%	7160.588	6001.276	5940.000	420
	DE1	4%	517.500	115.866	465.000	450
	DE2	9%	5476.667	6455.651	1830.000	60
	RS	66%	9104.015	5862.358	8700.500	22
	PSOGW	41%	16210.244	11193.375	15090.000	120
	PSOGC	45%	16818.667	12664.632	13800.000	630
	PSOLW	58%	18455.690	12870.897	14520.000	270
$N = 293 \times 307$	PSOLC	50%	16374.000	13597.782	13365.000	120
	DE1	7%	1598.571	1115.488	1470.000	120
	DE2	19%	17815.263	12484.580	16290.000	2730
	RS	64%	21548.531	13926.751	20852.500	57
	PSOGW	53%	31965.849	24423.975	27570.000	780
	PSOGC	45%	32532.667	22652.983	33210.000	1740
	PSOLW	55%	31472.182	23394.791	22620.000	720
$N = 397 \times 401$	PSOLC	54%	38156.111	22925.970	37665.000	750
	DE1	1%	1680.000	0.000	1680.000	1680
	DE2	12%	27722.500	17498.736	28620.000	180
	RS	60%	27302.567	21307.031	23607.500	145
	PSOGW	56%	49893.750	37515.327	44640.000	930
	PSOGC	55%	49975.636	36727.380	41760.000	300
	PSOLW	55%	49207.091	34053.904	50430.000	2010
$N = 499 \times 503$	PSOLC	46%	48443.478	34677.039	43470.000	1920
	DE1	1%	2480.000	0.000	2480.000	2480
	DE2	8%	67245.000	35114.316	64770.000	14730
	RS	61%	54139.443	38642.970	48743.000	140
	PSOGW	52%	72175.000	48653.823	71550.000	600
	PSOGC	51%	81476.471	53666.543	75100.000	5000
	PSOLW	49%	78651.020	48197.105	67400.000	11200
$N = 599 \times 601$	PSOLC	52%	69542.308	48837.949	53050.000	2500
	DE1	2%	4700.000	4808.326	4700.000	1300
	DE2	5%	8620.000	8078.180	9300.000	800
	RS	64%	86123.656	47504.284	89392.500	904
	PSOGW	46%	207443.478	163585.340	214800.000	800
	PSOGC	46%	175426.086	138118.794	149200.000	800
	PSOLW	60%	196993.334	146204.518	144500.000	9200
$N = 691 \times 701$	PSOLC	52%	209307.692	163833.606	200100.000	1800
	DE1	2%	23800.000	25000.000	23800.000	21000
	DE2	10%	71000.000	95357.642	15200.000	1600
	RS	60%	185932.334	126355.926	154999.000	2828

required a quite small number of function evaluations with respect to the cardinal of the domain of the function. Finally, it is important to note that in the experiments where the EC methods failed in obtaining a global minimizer, they located a local minimizer with value close to the global minimum.

Similar results are obtained for the minimization of the functions h_e (Problem 2) and w_e (Problem 3), and for $N = 103 \times 107$ they are reported in Table 1.2. In the case of Problem 3, the success rates of PSO method and its variants are high (around 80%) while the performance of DE variants remains low. However, random search again outperforms the EC methods in terms of success rates.

Table 1.2. Results for functions h_e (see (1.24)) and w_e (see (1.26)), for $N = 103 \times 107$

Function	Method	Suc.Rate	mean F.E.	St.D. F.E.	median F.E.	min F.E.
	PSOGW	51%	2013.333	1483.535	1500.000	100
	PSOGC	57%	1974.035	1609.228	1420.000	60
	PSOLW	59%	1677.288	1254.688	1420.000	60
h_e	PSOLC	58%	2385.862	1676.898	2040.000	120
	DE1	1%	100.000	0.000	100.000	100
	DE2	1%	80.000	0.000	80.000	80
	RS	65%	2099.646	1448.007	2056.000	6
	PSOGW	79%	1382.785	1265.927	820.000	40
	PSOGC	84%	1402.857	1442.194	930.000	40
	PSOLW	80%	1757.750	1544.267	1110.000	40
w_e	PSOLC	85%	1416.000	1329.034	880.000	40
	DE1	1%	60.000	0.000	60.000	60
	DE2	1%	80.000	0.000	80.000	80
	RS	96%	1507.969	1328.913	1104.000	7

1.4.2 Cryptanalysis of Feistel Ciphers through Evolutionary Computation Methods

In this section two different instances of a problem introduced by the Differential Cryptanalysis of a Feistel cryptosystem are considered and formulated as optimization tasks. Specifically, the problem of finding some missing bits of the key that is used in a simple Feistel cipher, namely the Data Encryption Standard with four and six rounds, respectively is studied [65, 66]. The two instances are complementary, since every problem of missing bits of keys in Differential Cryptanalysis of Feistel ciphers can be categorized into one of the two cases.

The performance of PSO and the DE on this problem is studied. Experimental results for DES reduced to four rounds show that the optimization methods considered, located the solution efficiently, as they required a smaller number of function evaluations compared to the brute force approach. For DES reduced to six rounds the effectiveness of the proposed algorithms depends on the construction of the objective function.

Problem Formulation

DES reduced to four rounds

For DES reduced to four rounds Differential Cryptanalysis (DC) uses a one–round characteristic occurring with probability 1, recovering at the first step of the cryptanalysis 42 bits of the subkey of the last round. Considering the case where the subkeys are calculated using the DES key scheduling algorithm, the 42 bits given by DC are actual key bits of the key and there are 14 key bits still missing for the completion of the key. The brute force attack (i.e. search among all 14-bit keys) requires testing 2^{14} trials. The right key should satisfy the known plaintext XOR value for all the pairs that are used by DC. An alternative approach is to use a second characteristic that corresponds to the missing bits and attempt a more careful counting on the key bits of the last two rounds, which is however more complicated.

Instead of using the aforementioned approaches to find the missing key bits, we formulate the problem of computing the missing bits as an integer optimization problem [71]. Since the right key should satisfy the known plaintext XOR value for all the pairs that are used by DC, these ciphertexts can be used for the evaluation of possible solutions provided by optimization methods. Thus, let $\mathbf{X}$ be a 14–dimensional vector, where each of its components corresponds to one of the 14 unknown key bits. Such a vector represents a possible solution of the optimization problem. Also, let np be the number of ciphertext pairs used by DC to obtain the right 42 key bits. Then we can construct the 56 bits of the key, using the 42 bits which are recovered by DC and the 14 components of $\mathbf{X}$ in the proper order. With the resulting key, decrypt the np ciphertext pairs and count the number of decrypted pairs that satisfy the known plaintext XOR value, denoted as $cnp_{\mathbf{X}}$. Thus, the objective function f is the difference between the desired output np and the actual output $cnp_{\mathbf{X}}$, i.e. $f(\mathbf{X}) = np - cnp_{\mathbf{X}}$. The global minimum of the function f is zero and the global minimizer is with high probability the actual key. A first study of this approach is given in [64].

DES reduced to six rounds

The cryptanalysis of DES reduced to six rounds is, as expected, more complicated than that of the four round version, since the best characteristic that can be used has probability less than 1. In particular, DC uses two characteristics of probability $p_{sr} = 1/16$ to provide 42 bits of the right key. Again, there are 14 bits of the key missing. However, in this case the right key may not be suggested by all ciphertext pairs. This happens because not all the corresponding plaintexts pairs are *right pairs*. A pair is called *right pair with respect to an r-round characteristic* $\Omega = (\Omega_P, \Omega_\Lambda, \Omega_C)$ *and an independent key* K, if it holds that $P' = \Omega_P$, where P' is the pair's XOR value, and for the first r rounds of the encryption of the pair using the independent key K the input and output XORs of the ith round are equal to λ_I^i and λ_O^i, respectively [6].

The probability that a pair with plaintext XOR equal to Ω_P of the characteristic is a right pair using a fixed key is approximately equal to the probability of the characteristic. A pair which is not a right pair is called *wrong pair* and it does not necessarily suggest the right key as a possible value. The study of right and wrong pairs has shown that the right key appears with the probability of the characteristic from the right pairs and some other random occurrences from wrong pairs. In conclusion, if all the pairs of DC (right and wrong) are used in the predefined objective function f, the function's minimum value will change depending on the specific pairs used. On the other hand, if the right pairs are filtered and are solely used in the objective function f, the function's global minimum will be constant with value equal to 0, as in the case of missing bits of DES reduced to four rounds. As the filtering of the right pairs is not always possible and easy, we study the behavior of the proposed approach using the objective function f with all the pairs of DC.

Experimental Setup and Results

Both the PSO and DE methods were applied considering each component of the possible solution as a real number in the range $[0, 1]$ and all populations were constrained to lie within the feasible region of the problem. For the evaluation of the suggested solutions, the technique of rounding off the real values of the solution to the nearest integer [71, 108] was applied. For the PSO method we have considered both the global and local variants, and for the DE algorithm all five variants described in Sect. 1.2. A maximum value for the velocity, $V_{\max} = 0.5$, of the PSO method was set in order to avoid the swarm's explosion, i.e. avoid velocities from assuming large values that lead to fluctuation of the particles over the search space, and thus destroy the dynamic of the method. The parameters of PSO were set at the default values, i.e. $\chi = 0.729$ and $c_1 = c_2 = 2.05$, found in the literature [21], and the parameters of DE were set at equal values $CR = F = 0.5$.

The proposed approach was tested for several different initial keys and number of pairs, np. For each setting, the size of each population was equal to 100 and the performance of the methods was investigated on 100 independent runs. A run is considered to be successful if the algorithm identifies the global minimizer within a prespecified number of function evaluations. The function evaluations threshold for both problems was taken equal to 2^{14}. For the missing bits of the key of DES reduced to four rounds, the results for six different keys, k_i, $i = 1, 2, \ldots, 6$, and for test pairs, np, equal to 20 and 50 are reported in Tables 1.3, 1.4, respectively. In the tables, PSOCG denotes the global variant of the constriction factor version of PSO, PSOCL1 denotes the PSO's local variant with neighborhood size equal to 1, PSOCL2 corresponds to PSO's local variant with neighborhood size equal to 2 (see Sect. 1.2), and DE1, DE2, DE3, DE4, DE5 denote the five DE variants of (1.5),(1.8),(1.9),(1.10),(1.11), respectively. Each table reports the success rate (Suc.Rate) of each algorithm, that is the proportion of the times it achieved the global minimizer within the prespecified threshold, and the mean value of function evaluations (Mean F.E.) over the successful experiments.

The results for the problem of recovering the missing (after the application of DC) bits of DES reduced to four rounds suggest that the proposed approach is able to locate the global minimizer i.e. the 14 missing bits of the key, with relatively low computational cost compared to the brute force attack. The success rates of all versions of the two methods are high. For np equal to 20 (Table 1.3) success rates range from 93% to 100%, with an average of 99.3%. For np equal to 50 (Table 1.4) the success rates lie in the region from 90% to 100%, with mean 99.4%. The improvement of the success rate as the number of ciphertext pairs, np, increases is expected as the larger number of ciphertext pairs used for the evaluation of the possible solutions reduces the possibility of a wrong 14-tuple being suggested as the right one. The mean number of function evaluations required to locate the global minimizer (over all variants) is 1309 in the case of $np = 20$ and 982 in the case of $np = 50$. This implies that, as more ciphertext pairs are incorporated in the objective function, not only the evaluation becomes more accurate, but also the global minimizer becomes easier to locate. However, the number of ciphertext pairs used by the proposed approach should not exceed the number of the ciphertext pairs that are used by DC for the initial problem, as this would increase the total cost of the cryptanalysis in terms of encryptions and decryptions.

With respect to the different variants of the PSO method, the local variant with neighborhood size 2 (PSOLC2) accomplished success rates close to 100%, in all instances of the first problem, with an average of 1489 function evaluations. The global variant of PSO (PSOGC) achieved success rates from 93% to 100% in different instances of the problem, but with an average of 898 function evaluations. This means that, although the global variant of PSO exhibits overall lower success rates, in the cases where both local and global variants of PSO are able to locate the minimizer, the global variant requires less function evaluations than the local variant.

The DE variants exhibited a stable and similar to each other behavior, with mean success rates 100% in all cases. One minor exception was DE4 that achieved mean success rates of 99% on two instances of the problem. DE1 required the lowest mean number of function evaluations (576) among all the considered methods and their variants.

For the missing bits of the key of DES reduced to six rounds, where both right and wrong pairs are used in construction of the objective function, the

Table 1.3. Results for DES reduced to four rounds for six different keys using $np = 20$ test pairs

key	Method	Suc.Rate	Mean F.E.
k_1	PSOGC	99%	742.42
	PSOLC1	100%	1773.00
	PSOLC2	100%	1255.00
	DE1	100%	614.00
	DE2	100%	1406.00
	DE3	100%	780.00
	DE4	100%	588.00
	DE5	100%	1425.00
k_2	PSOGC	99%	911.11
	PSOLC1	100%	2665.00
	PSOLC2	100%	1650.00
	DE1	100%	603.00
	DE2	100%	1518.00
	DE3	100%	879.00
	DE4	100%	615.00
	DE5	100%	1649.00
k_3	PSOGC	94%	1117.02
	PSOLC1	99%	2447.48
	PSOLC2	100%	1688.00
	DE1	99%	693.94
	DE2	100%	1497.00
	DE3	100%	805.00
	DE4	100%	690.00
	DE5	100%	1427.00
k_4	PSOGC	96%	876.04
	PSOLC1	100%	2089.00
	PSOLC2	100%	1418.00
	DE1	99%	701.01
	DE2	100%	1378.00
	DE3	100%	843.00
	DE4	100%	568.00
	DE5	100%	1362.00
k_5	PSOGC	97%	900.00
	PSOLC1	99%	1979.80
	PSOLC2	100%	1496.00
	DE1	100%	662.00
	DE2	100%	1493.00
	DE3	100%	848.00
	DE4	100%	662.00
	DE5	100%	1542.00
k_6	PSOGC	93%	1457.00
	PSOLC1	95%	4475.79
	PSOLC2	99%	2913.13
	DE1	100%	651.00
	DE2	100%	1717.00
	DE3	100%	1063.00
	DE4	99%	725.25
	DE5	100%	1583.00

Table 1.4. Results for DES reduced to four rounds for six different keys using $np = 50$ test pairs

key	Method	Suc.Rate	Mean F.E.
k_1	PSOGC	99%	860.61
	PSOLC1	100%	1698.00
	PSOLC2	100%	1141.00
	DE1	100%	485.00
	DE2	100%	1215.00
	DE3	100%	785.00
	DE4	100%	553.00
	DE5	100%	1382.00
k_2	PSOGC	94%	741.49
	PSOLC1	100%	1367.00
	PSOLC2	100%	1100.00
	DE1	99%	490.91
	DE2	100%	1081.00
	DE3	100%	669.00
	DE4	100%	521.00
	DE5	100%	1128.00
k_3	PSOGC	99%	631.31
	PSOLC1	100%	1217.00
	PSOLC2	100%	1035.00
	DE1	100%	385.00
	DE2	100%	1006.00
	DE3	100%	546.00
	DE4	100%	409.00
	DE5	100%	1016.00
k_4	PSOGC	90%	947.78
	PSOLC1	98%	2292.88
	PSOLC2	100%	1588.00
	DE1	98%	666.33
	DE2	100%	1342.00
	DE3	100%	838.00
	DE4	99%	649.50
	DE5	100%	1294.00
k_5	PSOGC	100%	707.00
	PSOLC1	100%	1763.00
	PSOLC2	100%	1193.00
	DE1	100%	445.00
	DE2	100%	1127.00
	DE3	100%	684.00
	DE4	100%	465.00
	DE5	100%	1131.00
k_6	PSOGC	96%	880.21
	PSOLC1	100%	2009.00
	PSOLC2	100%	1390.00
	DE1	100%	507.00
	DE2	100%	1250.00
	DE3	100%	692.00
	DE4	100%	563.00
	DE5	100%	1230.00

Table 1.5. Results for DES reduced to six rounds for six different keys using $np = 200$ test pairs

key	Method	Suc.Rate	Mean F.E.
	PSOGC	26%	7038.46
	PSOLC1	9%	2188.89
	PSOLC2	8%	3862.50
	DE1	36%	5191.67
k_1	DE2	52%	5515.39
	DE3	41%	5807.32
	DE4	51%	6364.71
	DE5	59%	6855.93
	PSOGC	24%	5037.50
	PSOLC1	3%	1500.00
	PSOLC2	7%	2357.14
	DE1	34%	6535.29
k_2	DE2	58%	6968.97
	DE3	40%	5945.00
	DE4	39%	6897.44
	DE5	61%	6932.79
	PSOGC	41%	4902.44
	PSOLC1	6%	4533.33
	PSOLC2	5%	7340.00
	DE1	48%	5070.83
k_3	DE2	61%	6967.21
	DE3	53%	6698.11
	DE4	48%	5889.58
	DE5	56%	7926.79
	PSOGC	47%	4912.77
	PSOLC1	13%	4407.69
	PSOLC2	23%	4134.78
	DE1	57%	6491.23
k_4	DE2	76%	7594.74
	DE3	66%	6418.18
	DE4	72%	5741.67
	DE5	76%	7001.32
	PSOGC	36%	5575.00
	PSOLC1	4%	1950.00
	PSOLC2	5%	4700.00
	DE1	51%	5688.24
k_5	DE2	62%	7803.23
	DE3	57%	5229.83
	DE4	53%	5377.36
	DE5	64%	6387.50
	PSOGC	37%	5624.32
	PSOLC1	5%	2920.00
	PSOLC2	9%	3377.78
	DE1	49%	5681.63
k_6	DE2	63%	7380.95
	DE3	50%	7048.00
	DE4	51%	5621.57
	DE5	64%	7679.69

results for the same six different keys tested for DES reduced to four rounds, and for test pairs, np, equal to 200, are reported in Table 1.5.

From Tables 1.3,1.4,1.5, we observe that there is a considerable difference between the success rates for the case of four rounds and the case of six rounds. This can be attributed to the fact that in the former case we work with a characteristic that occurs with probability 1 while in the latter case we work with a characteristic with smaller probability (1/16). This means that in the set of 200 ciphertext pairs used by the objective function, approximately 12 pairs are right and suggest the right tuple, while the remaining 188 pairs suggest tuples at random, thus decreasing, the possibility of suggestion of the right tuple. Consequently, since the objective function becomes more effective when more right pairs are available or equivalently, when the probability of the utilized characteristic is large, it is expected that in the four round case the performance of the methods should be better than in the six round case. Although, the wrong pairs used in the objective function of DES for six rounds are misleading for the evaluation of the right tuple of missing bits, the global variant of PSO (PSOGC) and all DE variants (DE1–DE5) were able to locate the missing bits on an average of 35% of independent runs for PSOGC and 55% for the DE variants over all six different keys tested (Table 1.5). The function evaluations required for the location of the right 14–tuple of missing bits in this case are on average 5600 for all methods.

Finally, an interesting observation from the results of the proposed approach is that in the case of DES reduced to four rounds all methods in independent runs were able to locate four different 14–tuples satisfying the condition criterion of the objective function. These four solutions of the problem differed in two fixed positions, the 10th and the 36th, of the DES key. In the case of DES reduced to six rounds just one solution, the right one, was located by all methods.

The results indicate that the proposed methodology is efficient in handling this type of problems, since on DES reduced to four rounds it managed to address the problem at hand using an average of 576 function evaluations in contrast with the brute force approach that requires $2^{14} = 16384$ evaluations. Furthermore, the results of DES reduced to six rounds suggest that the effectiveness of the proposed approach depends mainly on the construction of the objective function. This approach is also applicable to all Feistel cryptosystems that are amenable to differential cryptanalysis, thus motivating its use for other Feistel cryptosystems. Finally, as a future direction, we are interested in studying the effectiveness of the proposed approach not just for missing bits of the key produced by Differential Cryptanalysis but also for all the bits of the key of Feistel ciphers.

1.4.3 Utilizing Artificial Neural Networks to Address Cryptographic Problems

In this section we consider the Artificial Neural Networks approach and study its performance on some cryptographic problems [69]. Specifically, we study the approximation of the *Discrete Logarithm Problem* (DLP) and the *Diffie Hellman key–exchange protocol Problem* (DHP) over the finite field $\mathbb{Z}_p$, where p is prime, and the *factorization problem related to the RSA cryptosystem* [112] (all three problems are presented in Sect. 1.1.2).

Experimental Setup and Results

Training algorithms: In this study the ANN training algorithms considered were the Standard Back Propagation (BP) [114], the Back Propagation with Variable Stepsize (BPVS) [75], the Resilient Back Propagation (RPROP) [110], the On-Line Adaptive Back Propagation (OABP) [73] and the Scaled Conjugate Gradient (SCG) method [92]. All methods were extensively tested with a wide range of parameters. In most of the testing cases, the training methods did not exhibit significantly different performance, except from BP, which encountered difficulties in training most of the times.

Network architecture: Since the definition of an "optimal" network architecture for any particular problem is quite difficult and remains an open problem, we tested a variety of topologies with different numbers of hidden layers and with various numbers of neurons at each layer. The results reported are the best results obtained for each problem. The architecture used is described with a series of integers denoting the number of neurons at each layer.

Data normalization: To make the adaptation of the network easier, the data are transformed through the normalization procedure, that takes place before training. Assuming that the data presented to the network are in $\mathbb{Z}_p$, where p is a prime number, the space $S = [-1, 1]$, is split in p sub-spaces. Thus, numbers in the data set are transformed to analogous ones in the space S. At the same time, the network output is transformed to a number within $\mathbb{Z}_p$ using the inverse operation.

Network evaluation: For the evaluation of the network performance we first measured the percentage of the training data, for which the network was able to compute the exact target value. This measure is denoted by μ_0. However, as network output was restricted within the range $[-1, 1]$, very small differences in output, rendered the network unable to compute the exact target but rather to be very close to it. This fact resulted in the insufficiency of the μ_0 measure as a performance indicator. Thus we employed the $\mu_{\pm v}$ measure. This measure represents the percentage of the data for which the difference between desired and actual output does not exceed $\pm v$ of the real target.

We note that the "near" measure, $\mu_{\pm v}$, has different meaning for the DLP and the DHMP. The "near" $\mu_{\pm v}$ measure is very important in the case of the DLP. If the size of the $(\pm v)$ interval is $O\left(\log(p)\right)$ then the "near" measure can replace the "complete" μ_0 one. In general, for some small values of v the "near" measure is acceptable since the discrete logarithm computation can be verified i.e. computation of exponents over finite fields [104]. However, the verification of the Diffie–Hellman Mapping is an open problem. Sets of possible values for the Diffie–Hellman Mapping can be used to compute sets of possible values for the Diffie–Hellman key. The values of the Diffie–Hellman key can be tested in practice; they are symmetric keys of communication between the two users. The percentage of success for the "near" measure for DHMP can be compared with the corresponding percentage for the DLP. The results of the comparison can be related to the conjecture that the two problems are computationally equivalent.

In [84], both DLP and DHMP for several small prime numbers p have been tested. The input patterns were different values of the input variable of the discrete logarithm function and the Diffie–Hellman Mapping respectively, and the target patterns were the values of the corresponding function, for fixed chosen values of generators g and primes p. The ANNs in this case succeeded in training and generalizing, reaching up to 100%. Next, larger primes were tested rendering the task of training networks harder. Having so many numbers normalized in the range $[-1, 1]$ posed problems for the adaptation process. Thus, small changes in the network output caused complete failure, requiring the use of larger architectures, i.e., more nodes and layers. In cases with very large primes, the network performance on training was very poor. Some indicative results on training are reported in Table 1.6.

Table 1.6. Results for networks trained on the DLP and DHMP

p	Topology	Epochs	μ_0	$\mu_{\pm 2}$	$\mu_{\pm 5}$	$\mu_{\pm 10}$	Problem
83	$1 - 5 - 5 - 1$	20000	20%	30%	48%	70%	DLP
	$1 - 5 - 5 - 1$	20000	20%	35%	51%	70%	DHMP
97	$1 - 5 - 5 - 1$	25000	20%	30%	48%	70%	DLP
	$1 - 5 - 5 - 1$	20000	20%	35%	51%	70%	DHMP

The DLP is also studied in a different setting. More specifically, we have studied the case where, for several values of the prime p and the primitive root g, the value of $h = g^u (\bmod\ p)$, remains fixed. The input patterns consisted of pairs of primes p and the corresponding primitive roots g and the target patterns were the corresponding values of u, such that $\log_g h \equiv u (\bmod\ p)$, for a chosen fixed value h. We have tested the DLP in this setting for p assuming values between 101 and 2003, with several network topologies and training methods. In this case, there was a differentiation among the results obtained by different methods. For small values of p, i.e. from 101 to 199, the best results on the approximation of u, were obtained by the AOBP method. For larger values of p, the best results were given by the SCG method. Results on this new setting are reported in Table 1.7. All these results refer to training the ANNs on the approximation of the value of discrete logarithm u. Comparing the results exhibited in Tables 1.6 and 1.7, it seems that for the DLP problem, the approximation capability of the FNNs is better in the new setting.

Table 1.7. Results for networks trained on the second setting of the DLP

Range of p	Topology	Epochs	μ_0	$\mu_{\pm 15}$	$\mu_{\pm 20}$	$\mu_{\pm 30}$	$\mu_{\pm 40}$
$101 - 199$	$2 - 15 - 1$	600000	100%	100%	100%	100%	100%
$503 - 1009$	$2 - 25 - 1$	600000	82%	93%	96%	96%	98%
$1009 - 2003$	$2 - 30 - 1$	600000	17%	40%	46.7%	51.8%	54.1%
$1009 - 2003$	$2 - 3 - 3 - 3 - 1$	20000	7.5%	34.3%	44.8%	64.2%	71.6%

The ability of neural networks to address the RSA cryptosystem has also been investigated. In a previous work of ours, we have tried to approximate the $\phi(N)$ mapping, $N \mapsto \phi(N)$, with input patterns being numbers $N = p \times q$, where p and q are primes, and as target patterns the $\phi(N) = (p - 1) \times (q - 1)$ numbers. In this case the normalization problem was no more an obstacle. What is really interesting in this case is the generalization performance of the networks. Clearly, the networks were able not only to adapt to the training data, but also to achieve very good results with respect to the test sets [84]. Indicative results on networks trained for the $\phi(N)$ mapping are exhibited in Table 1.8.

Table 1.8. Results for networks trained for the $\phi(N)$ mapping with $N = p \times q \leqslant 10^4$

Topology	Epochs	μ_0	$\mu_{\pm 2}$	$\mu_{\pm 5}$	$\mu_{\pm 10}$	$\mu_{\pm 20}$
$1 - 5 - 5 - 1$	80000	3%	15%	35%	65%	90%
$1 - 7 - 8 - 1$	50000	6%	20%	50%	70%	100%

The factorization problem is also viewed in a different setting. More specifically, approximating the value of the function $p^2 + q^2$, given the value of N, leads directly to the factorization of N to its factors p and q. Thus, we tested the ANNs on the approximation of the aforementioned function for several instances of N. The results for this problem are reported in Table 1.9.

Table 1.9. Results for the second setting of the factorization problem for N ranging from 143 to 1003

Topology	Epochs	μ_0	$\mu_{\pm 15}$	$\mu_{\pm 20}$	$\mu_{\pm 30}$	$\mu_{\pm 40}$
$1 - 15 - 1$	200000	35.1%	36.8%	42.1%	43.8%	45.6%
$1 - 20 - 1$	600000	35.1%	43.8%	45.6%	52.6%	56.2%

Although the two settings of the factorization problem are computationally equivalent the approximation capabilities of the FNNs for this problem seem to be better for the first setting.

It is known that if a method for computing indices over finite fields is available, then the RSA cryptosystem breaks. In other words, the DLP is no easier than the factorization problem related to the RSA, which is confirmed by our experimental results.

In this study we consider only FNNs. In a future correspondence we intend to apply various other networks and learning techniques including non-monotone neural networks [11], probabilistic neural networks [122], self-organized maps [58], recurrent networks and radial basis function networks [41] among others. All data sets used in our experiments are available upon request.

1.4.4 Artificial Neural Networks Applied on Problems Related to Elliptic Curve Cryptography

In this section we study the performance of ANNs on the problem of computing the least significant bit of the discrete logarithm of a point over elliptic curves. The computation of the least significant bit of the discrete logarithm over elliptic curves with known odd order is important for cryptographic applications as it leads to the computation of all bits of the discrete logarithm. The results of this first attempt to address the specific problem using ANNs indicate that ANNs are able to adapt to the data presented with high accuracy, while the response of ANNs to unknown data is slightly higher than random selection. Another important finding is that ANNs require a small amount of storage for the known patterns in contrast to the storage needed for the data set itself [63].

Problem Formulation

For the discrete logarithm problem over elliptic curves, the following proposition is derived from the bit security of discrete logarithms over any cyclic group [101, 125].

Proposition 1. *Given an elliptic curve, E, over a finite field, $\mathbb{F}_q$, with known order n, and an oracle for a bit of the discrete logarithm that does not correspond to any power of 2 that divides the order n, then all the bits of the discrete logarithm can be computed in polynomial time.*

Remark 1. Currently, there is no polynomial algorithm for finding the order of an elliptic curve. Furthermore, the complexity for the computation of the discrete logarithm problem over elliptic curves with no knowledge of its order is exponential and, hence, it remains a computationally difficult task.

From Proposition 1 it is derived that in the case of an elliptic curve with odd order n, an oracle that gives the least significant bit of the discrete logarithm of a point over the elliptic curve leads to the computation of all bits of the discrete logarithm in polynomial time. Furthermore, prime order elliptic curves are considered more secure. Thus, our focus is on the computation of the least significant bit of the discrete logarithm of a point over elliptic curves of odd order. Complexity estimates for the computation of bits of the discrete logarithm over different fields can be found in [22, 120].

In relation to our problem, the considered Boolean function is defined as follows. Assume an elliptic curve $E(\mathbb{F}_p)$ and let $P = (x_P, y_P)$, $Q = (x_Q, y_Q)$ be two points of $E(\mathbb{F}_p)$, such that $Q = tP$, with $0 \leqslant t \leqslant (n-1)$. Define the Boolean function $f : \{0, 1\}^{4\lceil \log p \rceil} \mapsto \{0, 1\}$, with

$$f(x_P, y_P, x_Q, y_Q) = \text{lsb}(t), \tag{1.27}$$

which has inputs the coordinates x_P, y_P, x_Q, y_Q, in binary representation, and outputs the least significant bit (lsb) of t, i.e., 1 if the least significant bit of t is 1, and 0 otherwise. In general, a Boolean circuit that computes this function can be exponentially large in $\lceil \log p \rceil$ [22]. For the computation of this Boolean function we employ Artificial Neural Networks. Here, we focus on FNNs for the approximation of the Boolean function derived by elliptic curve cryptography defined in (1.27). For the general problem of the computation of a Boolean function by a FNNs, the following theorem proved in [2], supports the effectiveness of the proposed approach.

Theorem 2. *There is a threshold network with one hidden layer capable of computing any Boolean function.*

Experimental Setup and Results

Training ANNs with threshold units requires the use of training methods that do not employ information about the derivatives of the error function. Furthermore, as shown in [23], analog neural networks can be more powerful than neural networks using thresholds, even for the computation of Boolean functions. Thus, we study the performance of ANNs using the hyperbolic tangent activation function of (1.19), which approximates a threshold function as λ_2 tends to infinity. In all experiments the output layer consists of two neurons, and the neuron with the highest output value determines the class in which the computed bit is classified. Thus, if the first neuron's output value is smaller than the value of the second neuron, the bit is considered to belong to Class 0, which corresponds to a "0" value of the bit, and vice versa. This setting enables us to use training methods that employ derivatives of the error function. In particular, we have studied the performance of three training algorithms each from a different category of training algorithms, namely the Resilient Back Propagation method (RPROP) [110], the Adaptive On–line Back Propagation method (AOBP) [73] and the Differential Evolution algorithm (DE) [126]. Regarding the topology of the networks, we have tested a variety of topologies with various numbers of neurons at each layer. We report only the best results obtained for each problem.

For the construction of the datasets the ECC_LIB library for elliptic curve cryptography [59] was used. The performance of ANNs was tested for three different datasets of the considered Boolean function that correspond to randomly chosen p's of bit length 14, 20, and 32, respectively, where $\mathbb{F}_p$ is the finite field over which the elliptic curve is constructed. All data sets used are available upon request.

At each experiment the dataset was randomly partitioned into a training set and a test set. Two thirds of the dataset were assigned to the training set and the remaining one third comprised the test set. To evaluate the network performance, first we measure the average of the percentage of the training set over all 10 experiments, for which the network was able to correctly predict the least significant bit. Then, the network's performance is evaluated by measuring the average percentage of the test set over all experiments.

The best results, for the prescribed setting and $\lambda_2 = 1$, were obtained using the AOBP method and are reported in Tables 1.10, 1.11 and 1.12, respectively. The results indicate that for three bit lengths, ANNs are able to adapt to the training data with an average accuracy of 90%. With respect to the test sets, ANNs achieved for all three bit lengths an average accuracy of 57%, i.e. a slightly higher than random selection. Regarding the training epochs required in each case, as the bit length of p increases, more epochs are needed to achieve the same accuracy.

Another interesting finding regarding the training set, is that the network is able to learn the training patterns and respond correctly about the least significant bit of the discrete logarithm, using less storage than that required by the corresponding dataset. The results for the data compression are reported in Table 1.13. In Table 1.13, "BL(p)" denotes the bit length of p, "Data Stor." denotes the storage bits required for the dataset, "ANN Stor." denotes the storage bits required for the network weights and "Accuracy" corresponds to the accuracy of the network to identify the desired value for both classes.

An interesting line of further research is to study the performance of ANNs for larger values of p and elliptic curves of different order, as well as, in other related problems such as the computation of the order of elliptic curves.

1.5 Ridge Polynomial Networks for Cryptography

Ridge Polynomial Networks (RPNs) belong to the class of ANNs that are based on product type of neurons, i.e., neurons apply their activation function over the product of the weighted inputs. RPNs exhibit several advantages compared to ANNs based on summing units. The computation of the least significant bit of the discrete logarithm is important for cryptographic applications as it is related to the computation of all bits of the discrete logarithm [40]. For this reason, in this section we relate these two aspects providing some theoretical conclusions and insights for future research.

For completeness purposes let us first introduce the Pi-Sigma networks that are the building components of the RPNs, along with some theoretical background of RPNs. A Pi-Sigma network (PSN) is a feedforward network

Table 1.10. Results for p of bit length 14, using $56 - 3 - 2$ topology

Epochs		Train			Test		
		Class 0	Class 1	Accuracy	Class 0	Class 1	Accuracy
500	Class 0	168	33	83.58%	30	24	55.56%
	Class 1	48	151	75.88%	23	23	50.00%
650	Class 0	184	17	91.54%	33	21	61.11%
	Class 1	32	167	83.92%	23	23	50.00%
700	Class 0	183	18	91.04%	33	21	61.11%
	Class 1	30	169	84.92%	21	25	54.35%
1000	Class 0	186	15	92.54%	33	21	61.11%
	Class 1	25	174	87.44%	18	28	60.87%

Table 1.11. Results for p of bit length 20, using $80 - 3 - 2$ topology

Epochs		Train			Test		
		Class 0	Class 1	Accuracy	Class 0	Class 1	Accuracy
2000	Class 0	186	14	93.0%	32	26	55.17%
	Class 1	23	177	88.5%	17	25	59.52%
3000	Class 0	191	9	95.5%	30	28	51.72%
	Class 1	19	181	90.5%	21	21	50.00%
4000	Class 0	194	6	98.0%	32	26	55.17%
	Class 1	18	182	91.0%	19	23	54.76%
6000	Class 0	196	4	98.0%	33	25	56.90%
	Class 1	17	183	91.5%	20	22	52.38%

Table 1.12. Results for p of bit length 32, using $128 - 3 - 2$ topology

Epochs		Train			Test		
		Class 0	Class 1	Accuracy	Class 0	Class 1	Accuracy
4000	Class 0	193	5	97.47%	36	21	63.16%
	Class 1	16	186	92.08%	20	23	53.49%
5000	Class 0	193	5	97.47%	36	21	63.16%
	Class 1	15	187	92.57%	19	24	55.81%
8000	Class 0	193	5	97.47%	35	22	61.40%
	Class 1	14	188	93.07%	18	25	58.14%
9000	Class 0	193	5	97.47%	35	22	61.40%
	Class 1	14	188	93.07%	16	27	62.79%

Table 1.13. Data compression results

BL(p)	Data Stor.	ANN Stor.	Accuracy
14	23200	8400	89.99%
20	32800	11856	94.75%
32	52000	18768	95.27%

with a single "hidden" layer of linear units that uses product units in the output layer, i.e., it uses *products of sums* of input components. The presence of only one layer of adaptive weights at PSNs results in fast training. There are two types of PSNs, the *Analog Pi-Sigma Networks* (APSNs) and the *Binary Pi-Sigma Networks* (BPSNs). A generalization of APSN, the Ridge polynomial networks (RPNs) is proved to have universal approximation capability [37]. BPSNs, on the other hand, are capable of realizing any Boolean function [119].

In Fig. 1.4 a PSN with a single output is illustrated. This network is a fully connected two-layered feedforward network. However, the summing layer is not "hidden" as in the case of the Multilayer Perceptron (MLP), since the weights from this layer to the output layer are fixed at the value 1. This property contributes to reducing the required training time.

Let $\mathbf{x} = (1, x_1, \ldots, x_N)^\top$ be an $N+1$-dimensional augmented input column vector, where x_k denotes the k-th component of $\mathbf{x}$. The inputs are weighted by K $(N + 1)$-dimensional weight vectors $\mathbf{w}_j = (w_{0j}, w_{1j}, \ldots, w_{Nj})^\top$, $j = 1, 2, \ldots, K$ and summed by a layer of K linear summing units, where k is the desired order of the network. The output of the jth summing unit, h_j, is given as follows:

$$h_j = \mathbf{w}_j^\top \mathbf{x} = \sum_{k=1}^{N} w_{kj} x_k + w_{0j}, \quad j = 1, 2, \ldots, K. \tag{1.28}$$

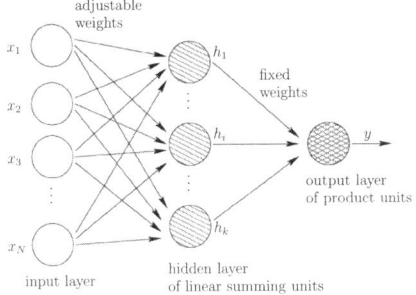

Fig. 1.4. A Pi-Sigma network (PSN) with one output unit

The output y of the network is given by:

$$y = \sigma(\prod_{j=1}^{K} h_j) = \sigma(net), \tag{1.29}$$

where $\sigma(\cdot)$ is a suitable activation function and $net = \prod_{j=1}^{K} h_j$. In the above, w_{kj} is an adjustable weight from input x_k to the jth summing unit and w_{0j} is the threshold of the jth summing unit. Weights can assume arbitrary real values.

The network shown in Fig. 1.4 is called a K-th order PSN since K summing units are incorporated. The total number of adjustable weight connections, including the adjustable thresholds, for a K-th order PSN with N-dimensional inputs is $(N+1)K$. If multiple outputs are required, an independent summing layer is needed for each output. Thus, for an M-dimensional output vector $\mathbf{y}$, a total of $\sum_{i=1}^{M}(N+1)K_i$ adjustable weight connections are present, where K_i is the number of summing units for the ith output. This enables the network to be incrementally expandable, since the order can be increased by adding another summing unit and associated weights, without disturbing any connection previously established. PSNs can handle both analog and binary input/output by using a suitable nonlinear activation function $\sigma(\cdot)$.

Regarding the approximation capabilities of PSNs, although the activation function is applied on a K-th order polynomial when K summing units are used and the exponents i_j sum up to K, this does not mean that only K-th order terms can be used, since by letting an extra input (the bias) be fixed at value 1, terms of order less than K are also realized. This K-th order polynomial, however, does not have full degrees of freedom since the coefficients are composed of sums and products of w_{kj}s and thus are not independent. Thus, a PSN cannot uniformly approximate all continuous multivariate functions that can be defined on a compact set. However, the theory of ridge polynomials can be used to show that universal approximation capability can be achieved simply by summing the outputs of APSNs of different orders. The resulting network is a generalization of PSN, which is called *Ridge Polynomial Network* (RPN), and is developed as follows.

For $\mathbf{x} = (x_1, \ldots, x_N)^\top$ and $\mathbf{w} = (w_1, \ldots, w_N)^\top \in \mathbb{R}^N$, we denote as $\langle \mathbf{x}, \mathbf{w} \rangle$ their inner product, i.e., $\langle \mathbf{x}, \mathbf{w} \rangle = \sum_{i=1}^N x_i w_i$. For a given compact set $C \subset \mathbb{R}^N$, all functions defined on C in the form of $f(\langle \mathbf{x}, \mathbf{w} \rangle)$, where f is a continuous function in one variable, are called ridge functions. A *ridge polynomial* is a ridge function that can be represented as

$$\sum_{i=0}^{n} \sum_{j=1}^{m} a_{ij} \langle \mathbf{x}, \mathbf{w} \rangle^i, \tag{1.30}$$

for some $a_{ij} \in \mathbb{R}$ and $\mathbf{w}_{ij} \in \mathbb{R}^N$.

It was proved in [15,16] that any polynomial in $\mathbb{R}^N$ with degree less than or equal to k can be represented by a ridge polynomial and, furthermore, it can be realized by an RPN [37]. From these results and the Stone-Weierstrass theorem the uniform approximation capability of the ridge polynomials of (1.30) is implied [37]. For applications of the RPNs in approximation and root finding see [31].

The ridge polynomial network (RPN), is defined as a feedforward network based on the generalized form of ridge polynomials [37]:

$$p(\mathbf{x}) = \sum_{j=1}^{n_{\text{total}}} \prod_{i=1}^{j} (\langle \mathbf{x}, \mathbf{w}_{ji} \rangle + w_{ij}), \tag{1.31}$$

where $n_{\text{total}} = \sum_{l=0}^{k} n_l$, and approximates an unknown function f on a compact set $C \subset \mathbb{R}^N$ as

$$f(\mathbf{x}) \approx (\langle \mathbf{x}, \mathbf{w}_{11} \rangle + w_{11}) + (\langle \mathbf{x}, \mathbf{w}_{21} \rangle + w_{21})(\langle \mathbf{x}, \mathbf{w}_{22} \rangle + w_{22}) + \cdots$$
$$+ (\langle \mathbf{x}, \mathbf{w}_{N1} \rangle + w_{N1}) \cdots (\langle \mathbf{x}, \mathbf{w}_{NN} \rangle + w_{NN}). \tag{1.32}$$

Each product term in (1.32) can be obtained as the output of a PSN with linear output units. Thus, the formulation of RPNs can be considered as a generalization of PSNs. Figure 1.5 represents a generic network architecture of the RPN using PSNs as building blocks. The RPN has only a single layer of adjustable weights which is beneficial in terms of training speed. Note that (1.32) serves as the basis for an incremental learning algorithm where PSNs of successively higher orders can be added until a desirable level of accuracy is obtained.

The uniform approximation capability of RPNs, their faster training compared to other kinds of ANNs, and the ability to perform incremental training, renders them a promising methodology for application to cryptological problems. Some conclusions derived by relating these two fields follow, providing an insight for future research.

Considering the computation of the rightmost bit of the discrete logarithm by real polynomials, the following theorem holds:

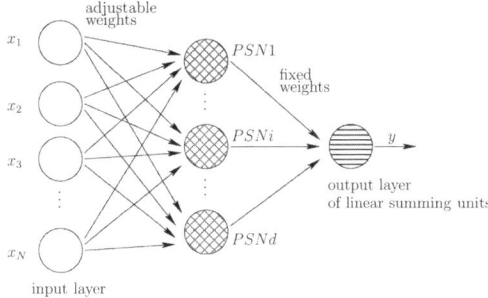

Fig. 1.5. A ridge polynomial network (RPN) with one linear output unit

Theorem 3 ([120]). *Let $0 \leqslant M < M + H \leqslant p - 1$. Assume that a polynomial $f(X) \in \mathbb{R}[X]$ is such that $f(x) \geqslant 0$ if x is a quadratic residue modulo p and $f(x) < 0$ otherwise, for every element $x \in S$ from some set $S \subseteq \{M + 1, \ldots, M + H\}$ of cardinality $|S| \geqslant H - s$. Then for any $\varepsilon > 0$ the bound*

$$\deg f \geqslant \begin{cases} H/2 - 2s - 1 - p^{1/2} \log p, & \text{for any } H, \\ \mathcal{C}(\varepsilon) \min\{H, H^2 p^{-1/2}\} - 2s - 1, & \text{if } p^{1/4+\varepsilon} \leqslant H \leqslant p^{1/2+\varepsilon}, \quad (1.33) \\ (p-1)/2 - 2s - 1, & \text{if } N = 0, H = p - 1, \end{cases}$$

holds, where $\mathcal{C}(\varepsilon) > 0$ depends only on ε.

Moreover, regarding real multivariate polynomials that compute the rightmost bit of the discrete logarithm, the following theorem holds:

Theorem 4 ([120]). *Let α_0, α_1 be two distinct real numbers, $r = \lfloor \log p \rfloor$, and let a polynomial $f(X_1, \ldots, X_r) \in \mathbb{R}[X_1, \ldots, X_r]$ be such that $f(\alpha_{u_1}, \ldots, \alpha_{u_r}) \geqslant 0$, if x is a quadratic residue modulo p and $f(\alpha_{u_1}, \ldots, \alpha_{u_r}) < 0$, otherwise, where $x = u_1 \ldots u_r$ is the bit representation of x, $1 \leqslant x \leqslant 2^r - 1$. Then f is of degree $\deg f \geqslant \log r + o(\log r)$ and contains at least $\operatorname{spr} f \geqslant 0.25r + o(r)$ distinct monomials.*

Since, the universal approximation capability of RPNs is achieved by summing the outputs of PSNs of different orders, for the approximation of the real polynomials that compute the rightmost bit of the discrete logarithm the following corollary is derived.

Corollary 1. *Assume the polynomial $f(X) \in \mathbb{R}[X]$ that satisfies the conditions of Theorem 3, i.e., it is such that $f(x) \geqslant 0$ if x is a quadratic residue modulo p and $f(x) < 0$ otherwise, for every element $x \in S$ from some set $S \subseteq \{M + 1, \ldots, M + H\}$ of cardinality $|S| \geqslant H - s$, and the polynomial $f(X_1, \ldots, X_r)$ that satisfies the conditions of Theorem 4, i.e., it is such that $f(\alpha_{u_1}, \ldots, \alpha_{u_r}) \geqslant 0$, if x is a quadratic residue modulo p and $f(\alpha_{u_1}, \ldots, \alpha_{u_r}) < 0$, otherwise, where $x = u_1 \ldots u_r$ is the bit representation of x, $1 \leqslant x \leqslant 2^r - 1$. Then, the following conclusions hold.*

(a) *There exist two RPNs that realize the polynomial $f(X)$ and the polynomial $f(X_1, \ldots, X_r)$, respectively, and*
(b) *The maximum orders of the nodes comprising each RPN, are required to be for the first case equal to*

$$\deg f(X) \geqslant \begin{cases} H/2 - 2s - 1 - p^{1/2} \log p, & \text{for any } H, \\ \mathcal{C}(\varepsilon) \min\{H, H^2 p^{-1/2}\} - 2s - 1, & \text{if } p^{1/4+\varepsilon} \leqslant H \leqslant p^{1/2+\varepsilon}, \\ (p-1)/2 - 2s - 1, & \text{if } N = 0, H = p - 1, \end{cases}$$
$$(1.34)$$

where $\mathcal{C}(\varepsilon) > 0$ depends only on ε, and equal to

$$\deg f(X_1, \ldots, X_r) \geqslant \log r + o(\log r), \qquad (1.35)$$

for the second case.

Additionally, the inherent ability of RPNs to expand the final polynomial that they realize after training, as an ordinary multivariate polynomial, may prove of major importance in the field of cryptography. Finally, relating the fact that many cryptographic results consider the use of Boolean functions with the capability of BPSNs to realize any Boolean function, can provide theoretical conclusions for this class of functions too.

1.6 Summary

The past decade has witnessed an increasing interest in the application of Computational Intelligence (CI) methods to problems derived from the field of cryptography and cryptanalysis. This is not only due to the effectiveness of these methods observed in many other scientific fields but also due to the need for automated techniques in the design and cryptanalysis of cryptosystems.

In this contribution, a brief review to cryptography and CI methods is initially provided. Then, a short survey of the applications of CI to cryptographic problems follows, and our contribution in this field is analytically presented. More specifically, at first three cryptographic problems derived from classical public key cryptosystems are formulated as discrete optimization tasks and Evolutionary Computation (EC) methods are applied to address them. Next, EC methods for the partial cryptanalysis of a Feistel cipher, the Data Encryption Standard reduced to four and six rounds, respectively, are considered. The effectiveness of Artificial Neural Networks (ANNs) for classical cryptographic problems and problems related to elliptic curve cryptography, follow. Lastly, some theoretical results are derived based on the composition of a specific class of ANNs, namely the Ridge Polynomial Networks, with theoretical issues of cryptography.

The experimental results presented for each considered problem suggest that problem formulation and representation are critical determinants of the performance of CI methods in cryptography. Regarding the application of EC methods in cryptanalysis, the proper definition of the fitness function such that no deceptive landscapes are created, is of major importance. Furthermore, the performance of ANNs in cryptographic problems depends on the problem formulation and data representation. A second conclusion derived is that EC methods (and CI methods in general) can be used as a practical assessment for the efficiency and the effectiveness of proposed cryptographic systems, meaning that they can "sense" flawed cryptographic schemes by finding patterns before more complex methods are employed for their analysis.

References

1. Adleman L (1979) A subexponential algorithm for discrete logarithm problem with applications to cryptography. In: Proceedings of the 20th FOCS, pp. 55–60
2. Anthony M (2003) Boolean functions and artificial neural networks. Technical report, CDAM, The London School of Economics and Political Science. CDAM Research Report LSE-CDAM-2003-01
3. Bäck T (1996) Evolutionary Algorithms in Theory and Practice : Evolution Strategies, Evolutionary Programming, Genetic Algorithms. Oxford University Press
4. Bagnall T, McKeown G. P, Rayward-Smith V. J (1997) The cryptanalysis of a three rotor machine using a genetic algorithm. In: Bäck T (ed) Proceedings of the 7th International Conference on Genetic Algorithms (ICGA97), San Francisco, CA, Morgan Kaufmann
5. Barbieri A, Cagnoni S, Colavolpe G (2004) A genetic approach for generating good linear block error-correcting codes. Lecture Notes in Computer Science 3103:1301–1302
6. Biham E, Shamir A (1991) Differential cryptanalysis of DES–like cryptosystems. Journal of Cryptology
7. Biham E, Shamir A (1993) Differential Cryptanalysis of the Data Encryption Standard. Springer–Verlag
8. Blake I (1999) Elliptic Curves in Cryptography. London Mathematical Society Lecture Notes Series vol. 265. Cambridge University Press
9. Blum A, Furst M, Kearns M, Lipton R. J (1994) Cryptographic primitives based on hard learning problems. Lecture Notes in Computer Science 773:278–291
10. Bonabeau E, Dorigo M, Théraulaz G (1999) From Natural to Artificial Swarm Intelligence. Oxford University Press, New York
11. Boutsinas B, Vrahatis M. N (2001) Artificial nonmonotonic neural networks. Artificial Intelligence 132:1–38
12. Burnett L, Carter G, Dawson E, Millan W (2001) Efficient methods for generating Mars-like S-boxes. Lecture Notes in Computer Science 1978(4):300–313
13. Carrol J, Martin S (1986) The automated cryptanalysis of substitution ciphers. Cryptologia 10(4):193–209
14. Chang Y.-C, Lu C.-J (2001) Oblivious polynomial evaluation and oblivious neural learning. Lecture Notes in Computer Science 2248:369–384

15. Chui C.K, Li X (1991) Realization of neural networks with one hidden layer. Technical report, Center for Approximation Theory, Dept. of Mathematics, Texas A&M University
16. Chui C.K, and Li X (1992) Approximation by Ridge functions and neural networks with one hidden layer. Journal of Approximation Theory 70:131–141
17. Clark A (1998) Optimisation Heuristics for Cryptography. PhD Thesis, Queensland University of Technology, Australia
18. Clark J.A, Jacob J.L (2000) Two-stage optimisation in the design of Boolean functions. Lecture Notes in Computer Science 1841:242–254
19. Clark J.A, Jacob J.L (2002) Fault injection and a timing channel on an analysis technique. Lecture Notes in Computer Science 2332:181–196
20. Clark J.A, Jacob J.L, Stepney S (2004) The design of S-boxes by Simulated Annealing. In: CEC 2004: International Conference on Evolutionary Computation, Portland OR, USA, June 2004, pp. 1517–1524. IEEE
21. Clerc M, Kennedy J (2002) The particle swarm–explosion, stability, and convergence in a multidimensional complex space. IEEE Transactions on Evolutionary Computation 6(1):58–73
22. Coppersmith D, Shparlinski I (2000) On polynomial approximation of the discrete logarithm and the Diffie–Hellman mapping. Journal of Cryptology 13:339–360
23. DasGupta B, Schnitger G (1996) Analog versus discrete neural networks. Neural Computation 8(4):805–818
24. De Jong K.A (1985) Genetic algorithms: A 10 year perspective. In: Proceedings of the First International Conference on Genetic Algorithms pp. 169–177. Lawrence Erlbaum Associates.
25. Diffie W, Hellman M.E (1976) New directions in cryptography. IEEE Transactions on Information Theory IT-22(6):644–654
26. Dontas K, Jong K (1990) Discovery of maximal distance codes using genetic algorithms. In: Proceedings of the Second International IEEE Conference on Tools for Artificial Intelligence pp. 805–811
27. Dorigo M, Gambardella L.M (1997) Ant colonies for the traveling salesman problem. BioSystems 43:73–81
28. Eberhart R.C, Simpson P, Dobbins R (1996) Computational Intelligence PC Tools. Academic Press
29. ElGamal T (1985) A public key cryptosystem and a signature scheme based on discrete logarithms. IEEE Transactions on Information Theory 31(4):469–472
30. Engelbrecht A (2002) Computational Intelligence: An Introduction. John Wiley & Sons
31. Epitropakis M.G, Vrahatis M.N (2005) Root finding and approximation approaches through neural networks. SIGSAM Bulletin: Communications in Computer Algebra, ACM Press 39(4):118–121
32. Feistel H (1973) Cryptography and computer privacy. Scientific American
33. Fogel D.B (1993) Evolving behaviours in the iterated prisoner's dilemma. Evolutionary Computation 1(1):77–97
34. Fogel D.B (1995) Evolutionary Computation: Towards a New Philosophy of Machine Intelligence. IEEE Press, Piscataway, NJ
35. Fogel D.B, Owens A.J, Walsh M.J (1966) Artificial Intelligence Through Simulated Evolution. John Wiley, Chichester, UK
36. Forsyth W.S, Safavi-Naini R (1993) Automated cryptanalysis of substitution ciphers. Cryptologia 17(4):407–418

37. Ghosh J, Shin Y (1992) Efficient higher-order neural networks for classification and function approximation. International Journal of Neural Systems 3:323–350
38. Goldberg D.E (1989) Genetic Algorithms in Search, Optimization, and Machine Learning. Addison Wesley, Reading, MA
39. Hassoun M.H (1995) Foundamentals of Artificial Neural Networks. MIT Press, Cambridge, MA
40. Håstad J, Näslund M (2004) The security of all RSA and discrete log bits. Journal of the ACM 51(2):187–230
41. Haykin S (1999) Neural Networks, A Comprehensive Foundation. Prentice Hall, New Jersey, USA, 2nd edition edition
42. Herdy M (1991) Application of the evolution strategy to discrete optimization problems. Lecture Notes in Computer Science pp. 188–192
43. Hernández J, Isasi P, Ribagorda A (2002) An application of genetic algorithms to the cryptoanalysis of one round TEA. In: Proc. of the 2002 Symposium on Artificial Intelligence and its Application
44. Hernández J, Sierra J, Isasi P, Ribagorda A (2002) Genetic cryptoanalysis of two rounds TEA. Lecture Notes in Computer Science 2331:1024–1031
45. Holland J.H (1975) Adaptation in Natural and Artificial Systems. University of Michigan Press, Ann Arbor
46. Hornik K (1989) Multilayer feedforward networks are universal approximators. Neural Networks 2:359–366
47. Hunter D, McKenzie A (1983) Experiments with relaxation algorithms for breaking simple substitution ciphers. The Computer Journal 26(1):68 71
48. Isasi P, Hernández J (2004) Introduction to the applications of evolutionary computation in computer security and cryptography. Computational Intelligence 20(3):445–449
49. Jakobsen T (1995) A fast method for cryptanalysis of substitution ciphers. Cryptologia 19(3):265–274
50. Karras D, Zorkadis V (2002) Strong pseudorandom bit sequence generators using neural network techniques and their evaluation for secure communications. Lecture Notes in Artificial Intelligence 2557:615–626
51. Karras D, Zorkadis V (2003) On neural network techniques in the secure management of communication systems through improving and quality assessing pseudorandom stream generators. Neural Networks 16:899–905
52. Kennedy J, Eberhart R.C (2001) Swarm Intelligence. Morgan Kaufmann Publishers
53. King J, Bahler D (1992) An implementation of probabilistic relaxation in the cryptanalysis of simple substitution ciphers. Cryptologia 16(3):215–225
54. Kinzel W, Kanter I (2002) Interacting neural networks and cryptography. In: Kramer B (ed) Advances in Solid State Physics. vol. 42, pp. 383–391. Springer–Verlag
55. Klimov A, Mityagin A, Shamir A (2002) Analysis of neural cryptography. Lecture Notes in Computer Science 2501:288–298
56. Knudsen L.R, Meier W (1999) A new identification scheme based on the perceptrons problem. Lecture Notes in Computer Science 1592:363–374
57. Koblitz N (1987) Elliptic curve cryptosystems. Mathematics of Computation 48:203–209
58. Kohonen T (2000) Self-Organizing Maps. Springer-Verlag, Berlin, 3rd edition

59. Konstantinou E, Stamatiou Y, Zaroliagis C (2002) A software library for elliptic curve cryptography. Lecture Notes in Computer Science 2461:625–637
60. Kotlarz P, Kotulski Z (2005) On application of neural networks for s-boxes design. Lecture Notes in Artificial Intelligence 3528:243–248
61. Koza J.R (1992) Genetic Programming: On the Programming of Computers by Means of Natural Selection. MIT Press, Cambridge, MA
62. Lange T, Winterhof A (2002) Incomplete character sums over finite fields and their application to the interpolation of the discrete logarithm by boolean functions. Acta Arithmetica 101(3):223–229
63. Laskari E.C, Meletiou G.C, Stamatiou Y.C, Tasoulis D.K, Vrahatis M.N (2006) Assessing the effectiveness of artificial neural networks on problems related to elliptic curve cryptography. Mathematical and Computer Modelling. to appear
64. Laskari E.C, Meletiou G.C, Stamatiou Y.C, Vrahatis M.N (2005) Evolutionary computation based cryptanalysis: A first study. Nonlinear Analysis: Theory, Methods and Applications 63:e823–e830
65. Laskari E.C, Meletiou G.C, Vrahatis M.N (2005) Problems of Cryptography as Discrete Optimization Tasks. Nonlinear Analysis: Theory, Methods and Applications 63:e831–e837
66. Laskari E.C, Meletiou G.C, Stamatiou Y.C, Vrahatis M.N (2006) Applying evolutionary computation methods for the cryptanalysis of Feistel ciphers. Applied Mathematics and Computation. to appear
67. Laskari E.C, Meletiou G.C, Tasoulis D.K, Vrahatis M.N (2005) Aitken and Neville inverse interpolation methods over finite fields. Applied Numerical Analysis and Computational Mathematics 2(1):100–107
68. Laskari E.C, Meletiou G.C, Tasoulis D.K, Vrahatis M.N (2005) Transformations of two cryptographic problems in terms of matrices. SIGSAM Bulletin: Communications in Computer Algebra, ACM Press 39(4):127–130
69. Laskari E.C, Meletiou G.C, Tasoulis D.K, Vrahatis M.N (2006) Studying the performance of artificial neural networks on problems related to cryptography. Nonlinear Analysis Series B: Real World Applications 7(5):937–942
70. Laskari E.C, Meletiou G.C, Vrahatis M.N (2004) The discrete logarithm problem as an optimization task: A first study. In: Proceedings of the IASTED International Conference on Artificial Intelligence and Applications pp. 1–6. ACTA Press
71. Laskari E.C, Parsopoulos K.E, Vrahatis M.N (2002) Particle swarm optimization for integer programming. In: Proceedings of the IEEE 2002 Congress on Evolutionary Computation pp. 1576–15812. IEEE Press.
72. Laskari E.C, Parsopoulos K.E, Vrahatis M.N (2002) Particle swarm optimization for minimax problems. In: Proceedings of the IEEE 2002 Congress on Evolutionary Computation pp. 1576–1581. IEEE Press.
73. Magoulas G.D, Plagianakos V.P, Vrahatis M.N (2001) Adaptive stepsize algorithms for on-line training of neural networks. Nonlinear Analysis T.M.A. 47(5):3425–3430
74. Magoulas G.D, Vrahatis M.N (2006) Adaptive algorithms for neural network supervised learning: a deterministic optimization approach. International Journal of Bifurcation and Chaos 16(7):1929–1950
75. Magoulas G.D, Vrahatis M.N, Androulakis G.S (1997) Effective backpropagation training with variable stepsize. Neural Networks 10(1):69–82

76. Magoulas G.D, Vrahatis M.N, Androulakis G.S (1999) Increasing the convergence rate of the error backpropagation algorithm by learning rate adaptation methods. Neural Computation 11(7):1769–1796
77. Mathews R (1993) The use of genetic algorithms in cryptanalysis. Cryptologia 17(4):187–201
78. Matsui M (1994) Linear cryptanalysis method for DES cipher. Lecture Notes in Computer Science 765:386–397
79. Matsui M, Yamagishi A (1992) new method for known plaintext attack of feal cipher. Lecture Notes in Computer Science pp. 81–91
80. Maurer U, Wolf S (1999) The relationship between breaking the diffie-hellman protocol and computing discrete logarithms. SIAM Journal on Computing 28:1689–1721
81. Meletiou G.C (1992) A polynomial representation for exponents in $\mathbb{Z}_p$. Bulletin of the Greek Mathematical Society 34:59–63
82. Meletiou G.C (1993) Explicit form for the discrete logarithm over the field $GF(p, k)$. Archivum Mathematicum (Brno) 29(1–2):25–28
83. Meletiou G.C, Mullen G.L (1992) A note on discrete logarithms in finite fields. Applicable Algebra in Engineering, Communication and Computing 3(1):75–79
84. Meletiou G.C, Tasoulis D.K, Vrahatis M.N (2003) Cryptography through interpolation approximation and computational inteligence methods. Bulletin of the Greek Mathematical Society 48:61–75
85. Menezes A, van Oorschot P, Vanstone S (1996) Handbook of applied cryptography. CRC Press series on discrete mathematics and its applications. CRC Press
86. Merkle R.C, Hellman M.E (1978) Hiding information and signatures in trapdoor knapsacks. IEEE Transactions on Information Theory 24:525–530
87. Michalewicz Z (1994) Genetic Algorithms + Data Structures = Evolution Programs. Springer, Berlin
88. Millan W, Clark A, Dawson E (1997) Smart hill climbing finds better boolean functions. In: Proceedings of the 4th Workshop on Selected Areas in Cryptography
89. Millan W, Clark A, Dawson E (1999) Boolean function design using hill climbing methods. Lecture Notes in Computer Science 1587:1–11
90. Miller V (1986) Uses of elliptic curves in cryptography. Lecture Notes in Computer Science 218:417–426
91. Mislovaty R, Perchenok Y, Kanter I, Kinzel W (2002) Secure key-exchange protocol with an absence of injective functions. Phys. Rev. E 66(6):066102-1–066102-5
92. Møller M.F (1993) A scaled conjugate gradient algorithm for fast supervised learning. Neural Networks 6:525-533
93. Mullen G.L, White D (1986) A polynomial representation for logarithms in $GF(q)$. Acta Arithmetica 47:255–261
94. National Bureau of Standards, U.S. Department of Commerce, FIPS pub. 46. Data Encryption Standard. January 1977.
95. Niederreiter H (1990) A short proof for explicit formulas for discrete logarithms in finite fields. Applicable Algebra in Engineering, Communication and Computing 1:55–57
96. Odlyzko A (2000) Discrete logarithms: The past and the future. Designs, Codes, and Cryptography 19(2–3):129–145

97. Parsopoulos K.E, Vrahatis M.N (2002) Initializing the particle swarm optimizer using the nonlinear simplex method. In: Grmela A, Mastorakis N.E (eds) Advances in Intelligent Systems, Fuzzy Systems, Evolutionary Computation pp. 216–221

98. Parsopoulos K.E, Vrahatis M.N (2002) Recent approaches to global optimization problems through particle swarm optimization. Natural Computing 1(2–3):235–306

99. Parsopoulos K.E, Vrahatis, M.N (2004) On the computation of all global minimizers through particle swarm optimization. IEEE Transactions on Evolutionary Computation 8(3):211–224

100. Peleg S, Rosenfeld A (1979) Breaking substitution ciphers using a relaxation algorithm. Communications of the ACM 22(11):598–605

101. Peralta R (1986) Simultaneous security of bits in the discrete log. Lecture Notes in Computer Science 219:62–72

102. Pincus A (1999) Approximation theory of the mlp model in neural networks. Acta Numerica pp. 143–195

103. Plagianakos V, Vrahatis M.N (2002) Parallel Evolutionary Training Algorithms for "Hardware-Friendly" Neural Networks. Natural Computing 1:307–322

104. Pohlig S.C, Hellman M (1978) An improved algorithm for computing logarithms over $GF(p)$ and its cryptographic significance. IEEE Transactions on Information Theory 24:106–110

105. Pointcheval D (1994) Neural networks and their cryptographic applications. In: Charpin P (ed) INRIA, Livres de resumes EUROCODE'94

106. Pointcheval D (1995) A new identification scheme based on the perceptrons problem. Lecture Notes in Computer Science 950:318–328

107. Ramzan Z (1998) On Using Neural Networks to Break Cryptosystems. PhD Thesis. Laboratory of Computer Science, MIT

108. Rao S.S (1996) Engineering Optimization–Theory and Practice. Wiley Eastern, New Delhi

109. Rechenberg I (1973) Evolutionsstrategie: Optimierung technischer Systeme nach Prinzipien der biologischen Evolution. Frommann-Holzboog Verlag, Stuttgart, Germany

110. Riedmiller M, Braun H (1993) A direct adaptive method for faster backpropagation learning: The RPROP algorithm. In: Proceedings of the IEEE International Conference on Neural Networks pp. 586–591

111. Rivest R (1991) Cryptography and machine learning. Lecture Notes in Computer Science 739:427–439

112. Rivest R, Shamir A, Adleman L (1978) A method for obtaining digital signatures and public key cryptosystems. Communications of the ACM 21:120–126

113. Rosen-Zvi M, Kanter I, Kinzel W (2002) Cryptography based on neural networks – analytical results. Journal of Physics A: Mathematical and General 35(47):L707–L713

114. Rumelhart D, Hinton G, Williams R (1986) Learning internal representations by error propagation. In: RumelhartD.E, McClelland J.L (eds) Parallel distributed processing: Explorations in the microstructure of cognition. vol. 1 pp. 318–362. MIT Press

115. Ruttor A, Kinzel W, Kanter I (2005) Neural cryptography with queries. Journal of Statistical Mechanics pp. P01009

116. Ruttor A, Kinzel W, Shacham L, Kanter I (2004) Neural cryptography with feedback. Physical Review E 69(4):046110–1–046110–7

117. Schwefel H.-P (1995) Evolution and Optimum Seeking. Wiley, New York
118. Shi Y, Eberhart R.C (1998) A modified particle swarm optimizer. In: Proceedings of the IEEE Conference on Evolutionary Computation. Anchorage, AK
119. Shin Y, Ghosh J (1991) Realization of Boolean functions using binary Pi-Sigma networks. In: Proceedings of the Conference on Artificial Neural Networks in Engineering. St. Louis
120. Shparlinski I (ed) (2003) Cryptographic Applications of Analytic Number Theory. Progress in Computer Science and Applied Logic. Birkhäuser Verlag
121. Silverman J.H (1986) The Arithmetic of Elliptic Curves. Springer-Verlag
122. Specht D.F (1990) Probabilistic neural networks. Neural Networks 3(1):109–118
123. Spillman R (1993) Cryptanalysis of knapsack ciphers using genetic algorithms. Cryptologia 17(4):367–377
124. Spillman R, Janssen M, Nelson B, Kepner M (1993) Use of a genetic algorithm in the cryptanalysis of simple substitution ciphers. Cryptologia 17(1):31–44
125. Stinson D (1995) Cryptography: Theory and Practice (Discrete Mathematics and Its Applications). CRC Press
126. Storn R, Price K (1997) Differential evolution–a simple and efficient heuristic for global optimization over continuous spaces. Journal of Global Optimization 11:341–359
127. Tasoulis D.K, Pavlidis N.G, Plagianakos V.P, Vrahatis M.N (2004) Parallel Differential Evolution. In: Proceedings of the IEEE 2004 Congress on Evolutionary Computation (CEC 2004), Portland
128. Terano T, Asai K, Sugeno M (1992) A Complete Introduction to the Field: Fuzzy Systems Theory and Its Applications. Academic Press
129. Vertan C, Geangala C (1996) Breaking the Merkle-Hellman cryptosystem by genetic algorithms: Locality versus performance. In: Zimmermann H, Negoita M, Dascalu D (eds), Real World Applications of Intelligent Technologies pp. 201–208. Editura Academiei Romanie, Bucharest
130. Vrahatis M.N, Androulakis G.S, Lambrinos J.N, Magoulas G.D (2000) A class of gradient unconstrained minimization algorithms with adaptive stepsize. Journal of Computational and Applied Mathematics 114(2):367–386
131. White H (1990) Connectionist nonparametric regression: Multilayer feedforward networks can learn arbitrary mappings. Neural Networks 3:535–549
132. Wilson D, Martinez T (1997) Improved heterogeneous distance functions. Journal of Artificial Intelligence Research 6:1–34
133. Winterhof A (2001) A note on the interpolation of the Diffie-Hellman mapping. Bulletin of Australian Mathematical Society 64(3):475–477
134. Winterhof A (2002) Polynomial interpolation of the discrete logarithm. Designs, Codes and Cryptography 25(1):63–72
135. Yue T.-W, Chiang S (2001) The general neural-network paradigm for visual cryptography. Lecture Notes in Computer Science 2084:196–206

2

Multimedia Content Protection Based on Chaotic Neural Networks

Shiguo Lian

SAMI Lab,France Telecom R&D Beijing
2 Science South Rd,Haidian District,
Beijing,100080,China
shiguo.lian@orange-ft.com

For chaotic neural network has both chaos' and neural network's properties, it is regarded as the most close one to human's thinking, and has been attracting more and more researchers since the past decade. The properties, such as parameter sensitivity, random similarity, learning ability, etc., make it suitable for information protection, such as data encryption, data authentication, intrusion detection, etc. For its simple structure generates complicated or random sequences, it has been used to design the encryption algorithms that are of lower time-cost than traditional computing-based ones, such as DES, IDEA or AES. These algorithms are suitable for multimedia encryption or authentication.

In this chapter, by investigating chaotic neural networks' properties, the low-cost cipher and hash function based on chaotic neural networks are proposed and used to encrypt and authenticate images or videos. First, this chapter gives a brief introduction to chaotic neural network based data encryption, including stream ciphers, block ciphers and hash functions. Secondly, the chapter analyzes chaotic neural networks' properties that are suitable for data encryption, such as parameter sensitivity, random similarity, diffusion property, confusion property, one-way property, etc. Thirdly, the chapter gives some proposals to design chaotic neural network based cipher or hash function, and uses these ciphers to construct media encryption and authentication methods. In media encryption, the stream cipher and partial encryption method are constructed, which are used to encrypt images or videos. In media authentication, the chaotic neural network based hash function and media authentication method are designed, which can detect the malicious tampering of the media content. The analyses and experiments show the practicability of these encryption algorithms. Finally, conclusions are drawn, and some open issues in this field are presented.

S. Lian: *Multimedia Content Protection Based on Chaotic Neural Networks*, Studies in Computational Intelligence (SCI) **57**, 51–78 (2007)
www.springerlink.com   © Springer-Verlag Berlin Heidelberg 2007

2.1 Introduction

In recent years, it is founded that there is chaos phenomenon in human's brain [1,2], and chaos theory can be used to explain some abnormal activities in the brain [3]. Thus, chaos dynamics provides a chance for the research in neural networks, such as using neural networks to generate chaos or constructing chaotic neural networks. After knowing the existence of chaos in human's brain, some scientists propose the theory of chaotic neural networks. The earliest work was done by Aibara [4] and Inoue et al [5,6] in 1990 and 1991. From then on, chaotic neural networks attract more and more researchers, and are used in more and more applications, such as algorithm optimization, pattern recognition and some other applications. Among these applications, information security is a challenging and interesting one.

Neural networks are used to design data protection schemes because of their complicated and time-varying structures [7]. For example, the ciphers [8,9] are constructed based on the random sequences generated from the neural networks. For the property of initial-value sensitivity, ergodicity or random similarity, chaos is also introduced to data protection. For example, the block cryptosystem is designed via iterating a chaotic map [10], and the one-way hash is constructed based on the chaotic map with changeable parameters [11]. As a combination of neural networks and chaos, chaotic neural networks (CNN) are expected to be more suitable for data encryption. For example, it is reported [12] that faster synchronization can be obtained by jointing neural network's synchronization and chaos' synchronization.

Due to both of the properties, chaotic neural networks are regarded more suitable for data encryption. Till now, various chaotic neural network based encryption algorithms have been reported, which can be classified into two classes: stream cipher and block cipher. The first one uses neural networks to construct stream ciphers, in which, neural network is used to produce pseudo-random sequences. Cauwenberghs [13] made use of VLSI to construct cell neurons that produce the random vectors suitable for data encryption. Chan et al [14] took advantage of Hopfield neural networks' chaotic property to produce pseudo-random sequence that is used to encrypt plaintexts. Karras et al [15] proposed the method to evaluate the property of the pseudorandom sequence generated from chaotic neural networks, and reported that the sequence generated from chaotic neural networks may be of better performance than the one generated from conventional methods. Caponetto et al [16] designed a secure communication scheme based on cell neural networks. These stream ciphers' security depends on the sequences' randomness.

The second one uses neural network to construct block ciphers, which makes use of chaotic neural networks' properties to encrypt plaintext block by block. Yue et al [17] constructed the image encryption algorithm based on chaotic neural networks, which encrypts several gray images into some binary images. Yen et al [18] used perception neurons to construct a block cipher, in which, the weight matrix and threshold vector are used as keys, and the neurons' input and output act as plaintext and ciphertext, respectively. Lian et al [19] proposed a cipher based on chaotic neural network, which is used to encrypt compressed images partially. Yee et al [20] designed a symmetric block cipher based on the multi-layer perception network. Generally, block ciphers' security depends on their computing security [21–23] (It is not practical from the viewpoint of computing complexity for attackers to break the cryptosystem in condition of not knowing the key) that is determined by the confusion and diffusion criterias.

With the development of computer technology and Internet technology, multimedia data (images, videos or audios) are widely used in human's daily life. Taking videos for example, in order to keep secure, some sensitive videos need to be protected. Till now, various encryption algorithms have been reported, such as DES, RSA, IDEA or AES. However, most of them are used in text/binary data not in video data, for videos are often of large volumes with real-time operations. These operations include browsing, displaying, cutting, copying or bit-rate control, etc. In order to meet these applications, new encryption algorithms should be studied. For chaotic signal's generation is often of lower cost than traditional pseudorandom sequence's generation, chaos based cipher has been used to encrypt such large-voluminal data as image or video. Fridrich [24] constructed the block cipher based on Baker map, Cat map or Standard map and used it to encrypt images, Chen et al [25] constructed the one based on 3-Dimensional Cat map and used it to encrypt real-time images, Yen et al [17] designed the image encryption algorithm based on chaotic neural network, Lian et al [19] proposed the chaotic neural network based cipher to encrypt images or videos, and Li et al [26] proposed a chaos-based video encryption scheme (CVES) and used it to encrypt videos. These chaotic ciphers are more efficient compared with traditional ones.

In this chapter, the chaotic neural network's properties suitable for data encryption are investigated, and the design of efficient stream cipher and hash function based on chaotic neural networks are presented. Based on these ciphers, the secure and efficient media encryption and authentication schemes are proposed. The analyses and experiments show the practicability of the schemes. The rest of the chapter is arranged as follows. In Section 2, the chaotic neural networks' generation is presented, together with their cipher-suitable properties, such as diffusion, one-way, parallelization, parameter sensitivity or random similarity, etc. The media encryption methods based on chaotic neural networks are proposed in Section 3, including a brief introduction to media encryption, the design of chaotic neural networks based stream cipher, the design of the partial encryption methods for compressed images or videos, and some experimental results on the proposed encryption schemes. Similarly, in Section 4, the media authentication scheme based on chaotic neural networks is presented, including a brief introduction to media authentication, the design of hash function based on chaotic neural networks, the design of the authentication method, and the experiments and analyses.

Then, some open issues and future work are discussed in Section 5. Finally, in Section 6, some conclusions are drawn.

2.2 Chaotic neural networks' generation and properties

2.2.1 Chaotic neural network's generation

Generally, three methods are used to construct chaotic neural networks: 1) adopt neural networks' learning ability to approach a chaotic dynamic system; 2) analyze the condition that makes a neural network generate chaos; 3) design the chaotic neuron model and then the corresponding chaotic neural network.

In the first method, the chaos system is known, and the problem is to find a neural network that approaches the chaos system with it confirmed that the performance of the neural network is similar with the one of the chaos system. Thus, the synchronized neural network is a chaotic neural network. For example, Ren et al [27] used Elman network [28] to approach Hennon chaotic system, and the synchronization algorithm is based on feedback chaos control theory [29]. Generally, this method's difficulty is to realize chaos synchronization [30,31].

In the second method, the neural network is known, and the problem is to find the condition, under which, the neural network generates chaos. Among the existing neural networks, Hopfield network attracts the most researchers. The basic idea is to introduce chaos dynamics to Hopfield network. For example, Zou et al [32] found chaotic attractors in the dynamic activity of two cells' cellular neural networks, and they [33] reported that 3-order autonomous cellular neural network has chaotic attractors. Generally, the nonlinear property of neural network determines that it itself may have chaotic property, while the difficulty is to find the suitable condition.

The third method is to construct the chaotic neural network by designing a chaotic neuron model. The well-known examples include the Global Coupling Map (GCM) model [34], Aihara's chaotic neural network model [35] and Inoue's chaotic neural network model [36]. Among them, the GCM model proposed by Kaneko is composed of many neurons, each of which satisfies chaotic Logistic map [37]. Aihara's model is composed of some neurons, and each neuron includes the feedback coming from the internal neurons. Inoue's model is composed of some neurons, each of which is a coupling chaotic oscillator. There are some other methods to design chaotic neurons, such as the one using network model and Lorenz equation [38] and the one using a four-level feedback neural network [39].

Among the mentioned methods, the third one is more practical than the others. For the first one, the constructed chaotic neural network acts just as the original chaos system. For the second one, the process to find the chaotic dynamic is time cost and it limits the number of the chaotic neural network. The third one gives a general method to construct chaotic neural networks, which is easy to produce the chaotic neural network with both the chaos system's property and the neural network's property. Thus, the third method is considered in the following content.

2.2.2 Chaotic neural network's properties suitable for data encryption

Chaotic neural networks have both chaos' properties and neural networks' properties, such as diffusion, one-way property, parallel implementation, confusion, parameter sensitivity and randomness similarity.

Diffusion property

The neuron layer of a neural network has the property of diffusion. Taking the simple neuron layer in Fig. 2.1 for example, the neuron layer is composed of n neurons, f is a linear function, P is the input, and C is the output. It is defined as

$$C = \begin{bmatrix} c_0 \\ c_1 \\ \vdots \\ c_{n-1} \end{bmatrix} = f(\begin{bmatrix} w_{0,0} & w_{0,1} & \cdots & w_{0,n-1} \\ w_{1,0} & w_{1,1} & \cdots & w_{1,n-1} \\ \vdots & \vdots & \vdots & \vdots \\ w_{n-1,0} & w_{n-1,1} & \cdots & w_{n-1,n-1} \end{bmatrix} \begin{bmatrix} p_0 \\ p_1 \\ \vdots \\ p_{n-1} \end{bmatrix} + \begin{bmatrix} b_0 \\ b_1 \\ \vdots \\ b_{n-1} \end{bmatrix})$$

where

$$c_i = f(\sum_{j=0}^{n-1} w_{i,j} p_j + b_i) = \sum_{j=0}^{n-1} w_{i,j} p_j + b_i (i = 0, 1, \cdots, n-1)$$

Thus, each element c_i $(i = 0, 1, \cdots, n-1)$ in the output C is in relation with all the elements p_j $(j = 0, 1, \cdots, n-1)$ in the input P. That is, the change of one element in P may cause the changes of all the elements in C. This property is similar to the diffusion criteria [22] of the cipher with high computational security. In the diffusion criteria, the slight change in the plaintext or the key causes great changes in the ciphertext. The diffusion property of neural network makes it a potential choice for cipher designing.

One-way property

Some neural networks have the one-way property. Fig. 2.2 shows a simple neuron model with one-way property. In this neuron, the input is composed of n elements $p_0, p_1, \cdots, p_{n-1}$, while the output is a unique element c. It is defined as

$$c = f(\sum_{j=0}^{n-1} w_j p_j + b)$$

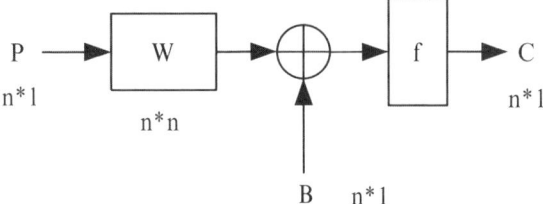

Fig. 2.1. A simple neuron layer with diffusion property

As can be seen, it is easy to compute c from $P = [p_0, p_1, \cdots, p_{n-1}]$, while difficult to compute P from c. The difficulty is equal to solve a singular equation. Thus, it is a one-way process from the input P to the output c. The one-way property makes neural network a suitable chaotic for hash function. As a hash function, it should be easy to compute the hash value from the plaintext, while difficult to compute the plaintext from the hash value. The hash value is often much shorter than the plaintext, which makes it suitable for digital signature or data authentication. According to this case, neural network may be used in hash function.

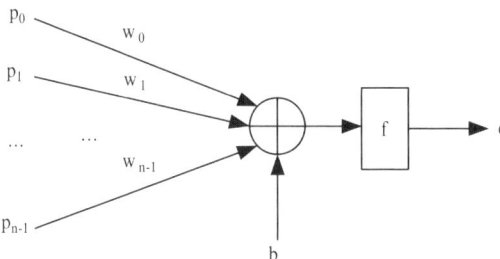

Fig. 2.2. A simple neuron layer with one-way property

Parallel implementation

Parallel implementation is an important property of neural networks, which benefits from neural networks' structure. Generally, in each layer of a neural network, the neurons are paralleled. Thus, each neuron can implement certain functionality independently. This property makes it suitable for parallel data processing. Fig. 2.3 shows the traditional data processing scheme (a) and the neural network based scheme (b). In (a), the data blocks $p_0, p_1, \cdots, p_{n-1}$ are operated one by one in order, which is named waterline scheme. In (b), the data blocks $p_0, p_1, \cdots, p_{n-1}$ are operated in a parallel way, which is named parallel scheme. Generally, the parallel scheme (b) saves much time cost than the waterline scheme (a).

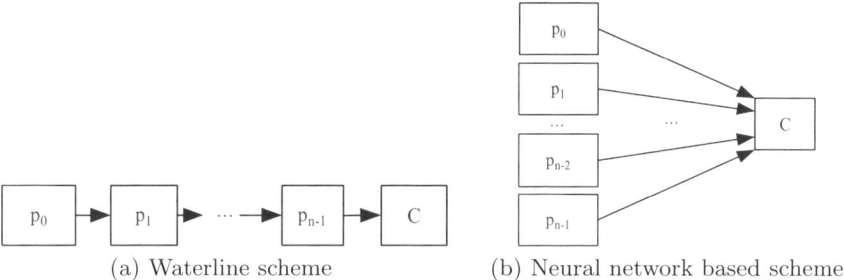

(a) Waterline scheme   (b) Neural network based scheme

Fig. 2.3. Comparison of data block processing schemes

Confusion

Confusion is a special property caused by the nonlinear structure of chaos systems or chaotic neural networks. The ergodicity and randomness of chaotic dynamic make the output depend on the input in a nonlinear and complicated manner [37]. Fig. 2.4 shows the piecewise linear chaotic map [37]. Here, x is the input, $g()$ is the chaotic map, and $y = g(x)$ is the output. Thus, for each output y, there exists 4 corresponding inputs x. Furthermore, if the chaotic map is iterated for n times, $y = g^n(x)$, then each output y corresponds to 4^n inputs x. Thus, it is difficult to determine the exact input. This property is similar to the confusion criteria of the block cipher with high computational security [22]. In the confusion criteria, the bit of ciphertext depends on all the bits of the plaintext and the key in a complicated way. The confusion property of chaos or chaotic neural networks makes them a potential choice for cipher designing.

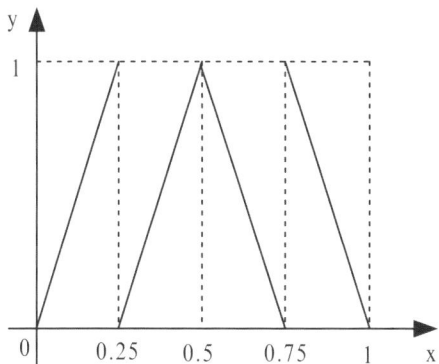

Fig. 2.4. The piecewise linear chaotic map

Parameter sensitivity

Chaos systems often have high parameter sensitivity, known as butterfly effect [37]. That is, the slight change in the original parameter will causes great changes in the final one. Here, the parameter can be either the initial value or the control parameter of the chaotic map. Taking the initial value for example, the high sensitivity benefits from the positive Lyapunov exponent of the chaos system. In a chaos system, Lyapunov exponent λ shows the bifurcation property of the chaotic orbit. It is defined as

$$\Delta Y = e^{\lambda t}|\Delta X| = (e^\lambda)^t|\Delta X|,$$

where ΔX is the slight difference in the initial value, and ΔY is the difference after t times of iteration. For a chaos system, $\lambda > 0$ is confirmed. Thus, the difference ΔY increases with t. Fig. 2.5 shows an example of chaotic Logistic map with $\Delta X = 0.000001$ and $t = 50$. As can be seen, the two sequences are close when $t < 20$, while the iterated difference is $\Delta Y = 0.5975$ that is much larger than ΔX when $t = 50$. The parameter sensitivity is similar to ciphers' key sensitivity. That is, a slight change in the decryption key causes great changes in the decrypted data. Thus, the sensitive parameter of chaotic maps can act as the encryption key in data encryption.

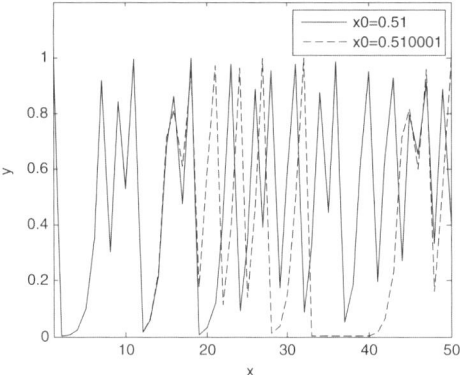

Fig. 2.5. The initial-value sensitivity of the chaotic Logistic map

Randomness Similarity

For chaos systems or chaotic neural networks, the produced sequence is often of random distribution. This property benefits from the property of ergodicity. That is, the chaotic sequence experiences all the states in the chaotic space.

Taking the chaotic Logistic map for example, the chaotic sequence x_i experiences all the value in the section $[0,1]$. Additionally, the chaotic sequence has some pseudorandom properties in autocorrelation function, cross-correlation function or frequency statistics. Fig. 2.6 shows an example on the chaotic Logistic map with $x_0 = 0.1$ and $x_1 = 0.11$. As can be seen, the x_0-sequence's autocorrelation (a) is similar to impulse function, and the cross-correlation (b) between x_0-sequence and x_1-sequence is near to 0. These properties make the chaos system a choice for pseudorandom sequence generation.

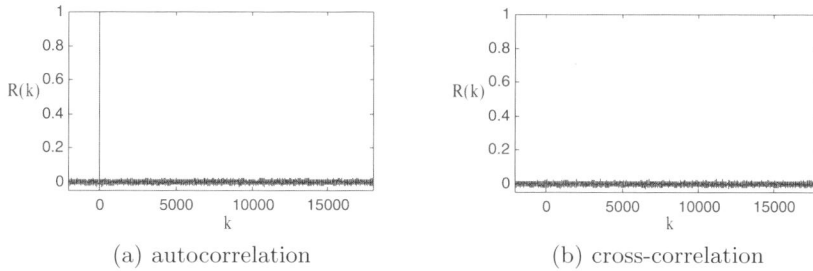

(a) autocorrelation   (b) cross-correlation

Fig. 2.6. The statistical results of chaotic sequence

2.3 Multimedia content encryption based on chaotic neural networks

2.3.1 Introduction to multimedia content encryption

In the past decade, some image or video encryption algorithms have been reported, which can be classified into two categories: complete encryption and partial encryption. The former one encrypts the raw data or the compressed data directly with traditional or chaotic ciphers. This category can be further divided into two types. The first one is to permute raw images or videos directly [40–42], which reduce media data's understandability by changing adjacent pixels' relationship. For example, Maniccam et al [42] proposed an algorithm that permutes images or videos with SCAN patterns. These algorithms are often used to encrypt TV signals that are not compressed. The second one is to encrypt the compressed data directly [43, 44]. For example, Qiao et al [43] proposed Video Encryption Algorithm (VEA) that encrypts only half of the plaintext with DES and decreases the time cost to 46%. These algorithms are often of high security, but also of high complexity, and often change the file format. According to these properties, complete encryption algorithm is more suitable for secure video storing than for real-time transmission.

Compared with complete encryption, partial encryption incorporates encryption into compression and encrypts images or videos partially. This is based on the assumption that an encryption scheme is regarded as secure if the cost for breaking it is not smaller than the one paid for the media data. Some algorithms encrypt only the DCT coefficients or motion vectors in the compressed media data [45–47]. For example, Shi et al [45] proposed the algorithms that encrypt only the signs of DCT coefficients, and Yen et al [47] proposed the one that encrypts motion vectors with the Chaotic Pseudo-Random Sequence Generator (CPRSG). Some other algorithms permute DCT coefficients partially or completely during compression [48–51]. For example, Tosum et al [48] proposed the one that permutes 64 coefficients in three sub-bands in order to reduce the effect on compression ratio, and Qiao et al [49] analyzed these permutation algorithms and told that they are not secure enough against known-plaintext attacks. Some other algorithms incorporate encryption into Variable Length Code (VLC). For example, Wu and Kuo [52, 53] proposed Multi-Huffman Trees (MHT) and Multi-State Indexes (MSI) algorithms. These partial encryption algorithms often meet real-time requirement and keep file format unchanged. For these features, they are more suitable for such real-time applications as video transmission or video access.

According to the above analyses, partial encryption is preferred in order to obtain good tradeoff between security, real-time operation and format compliance [19,54,55]. Some media encryption algorithms based on chaos or chaotic neural networks have been presented. However, most of them belong to complete encryption. For example, Fridrich [24] and Chen et al [25] proposed the block cipher based on 2-D chaotic map to encrypt uncompressed images, Yen et al [17] proposed the chaotic neural network based scheme to encrypt uncompressed images, and Li et al [26] proposed the chaos based block cipher to encrypt compressed videos. These algorithms cannot keep the synchronization information in the file format. According to this case, the partial encryption algorithm based on chaotic neural networks will be presented in the following content. Considering that the encrypted part is often of variable size, the adopted cipher should support the plaintext with variable size. Thus, the chaotic neural network based cipher will be designed, that is not only efficient but also supports variable-size plaintext.

2.3.2 The cipher based on chaotic neural network

For the random similarity of chaotic neural networks, they are suitable for pseudorandom number generation. Here, we propose a stream cipher based on a chaotic neural network, which is a modification of the one developed in [18]. The new cipher proposed here is described as follows.

The encryption process

The encryption process is synthesised in (2.1). Here, $f(x)$ is a function taking value 1 if $x \geq 0$ and 0 otherwise, P and C are the plaintext and the ciphertext, respectively. The weight ω and θ are determined by encryption/decryption keys, which will be presented in the following content. P is composed of N bits, and $P = [p_0, p_1, \cdots, p_{N-1}]^T (p_i = 0 \ or \ 1)$. Similarly, $C = [c_0, c_1, \cdots, c_{N-1}]^T (c_i = 0 \ or \ 1)$, $\theta = [\theta_0, \theta_1, \cdots, \theta_{N-1}]^T (\theta_i = \pm 1/2)$, and

$$\omega = \begin{bmatrix} \omega_{0,0} & 0 & \cdots & 0 \\ 0 & \omega_{1,1} & \ddots & \vdots \\ \vdots & \ddots & \ddots & 0 \\ 0 & \cdots & 0 & \omega_{N-1,N-1} \end{bmatrix} \quad (\omega_{i,i} = \pm 1, i = 0, \cdots, N-1).$$

$$C = f(\omega P + \theta) \tag{2.1}$$

The decryption process is symmetric to the encryption process. In decryption, the plaintext P is replaced by the ciphertext C, and the ciphertext C by the plaintext P. In the proposed cipher, the size of the plaintext, N, is arbitrary.

Generation of the parameters

The parameters ω and θ are generated from a chaotic sequence generator, as shown in Fig. 2.7. Here, the chaotic sequence X is firstly generated, then quantized and finally used to decide the two parameters. The process is described in detail as below.

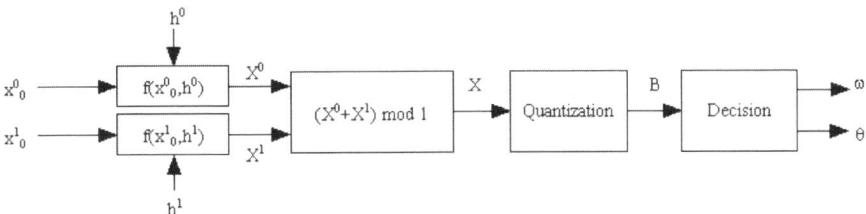

Fig. 2.7. Parameter generation of the chaotic neural network

Firstly, the chaotic skew tent map, shown in Eq. 2.2, is used to generate a chaotic sequence $X = x_0, x_1, \cdots, x_{N-1}$.

$$x_{j+1} = f(x_j, h) = \begin{cases} x_j/h & 0 < x_j \leq h \\ (1 - x_j)/(1 - h) & h < x_j \leq 1 \end{cases} \quad (j = 0, 1, \cdots, N-1) \tag{2.2}$$

Here, the parameters h and x_0 both range from 0 to 1, which are used for the key. According to the parameter sensitivity, the chaotic map is iterated for more than 50 times, and the 51-th iteration corresponds to x_0, 52-th iteration corresponds to x_1, and so on. Considering that the word length of computer

implementation is limited, the computing process may degrade the chaos'
properties [56]. In order to reduce the degradation, two sequences X^0 and X^1
and compounded together, which produces X. Among them, X^0 is produced
by the parameter pair (x_0^0, h^0), and X^1 by (x_0^1, h^1). And the compounding
operation is

$$(x_i^0 + x_i^1) mod 1 = \begin{cases} x_i^0 + x_i^1, & 0 < x_i^0 + x_i^1 \leq 1 \\ x_i^0 + x_i^1 - 1, & 1 < x_i^0 + x_i^1 \leq 2 \end{cases} \ (i = 0, 1, \cdots, N-1)$$

Then, the chaotic sequence X is quantized, which produces the chaotic binary
sequence B=$b_0, b_1, \cdots, b_{N-1}$. A simple example of quantization is

$$b_i = \begin{cases} 1, x_i \leq 0.5 \\ 0, else \end{cases}$$

Finally, the chaotic binary sequence B is used to determine the parameters
ω and θ, as shown below. For example, if $b_i = 0$, then $\omega_{i,i} = -1$, otherwise,
$\omega_{i,i} = 1$. Similarly, if $b_i = 0$, then $\theta_i = -1/2$, otherwise, $\theta_i = 1/2$.

$$\omega_{i,i} = \begin{cases} -1, b_i = 0 \\ 1, \quad b_i = 1 \end{cases}$$

$$\theta_i = \begin{cases} -1/2, b_i = 0 \\ 1/2, \quad b_i = 1 \end{cases}$$

The chained encryption mode

A chained encryption mode is used here to enhance the cryptosystem's security. The encryption process is

$$C_i = E(P_{i-1} \oplus P_i, K_i)$$

Here, P_i, P_{i-1}, K_i are i-th plaintext, i-1-th plaintext and i-th key respectively,
$\oplus$ means the bitwise operation, and $E()$ is the encryption algorithm. The
decryption process is

$$P_i = D(C_i, K_i) \oplus P_{i-1} = D(C_i, K_i) \oplus D(C_{i-1}, K_{i-1}) \oplus P_{i-2}$$
$$= D(C_i, K_i) \oplus D(C_{i-1}, K_{i-1}) \oplus \cdots \oplus D(C_1, K_1) \oplus D(C_0, K_0) \oplus P_{-1}$$

Here, $D()$ is the decryption algorithm. Seen from the equation, the i-th
ciphertext depends on all the previous ones, but is not affected by the following ones. Thus, the encrypted data stream can be directly cut off at the end,
which makes it suitable for applications with the bit-rate control requirement.

Performances of the proposed stream cipher

In this cipher, for simplicity, the perception neuron is used. Some other networks can also be used to construct the stream cipher, which will cause higher computing cost. Additionally, compared with the ciphers proposed in [13–17], this cipher is composed of fixed computing instead of floating computing, which makes the cryptosystem independent of the computer's precision. Compared with the one proposed in [18], this cipher contains a chaotic pseudorandom sequence generator with good properties, adopts the chained encryption mode that enhances the system security, and supports the plaintext with variable size.

The randomness of the chaotic pseudorandom sequence is tested [57] under the condition of $x_0^0 = 0.1, h^0 = 0.3, x_0^1 = 0.6$ and $h^1 = 0.8$. The results are shown in Table 2.1. As can be seen, the produced binary sequence is of good randomness. In the key of the stream cipher, the four parameters are all of high sensitivity that benefits from the chaotic map's properties mentioned above. Fig. 2.8 gives the test results of the cross-correlation between the plaintext and the ciphertext, and of the one between different ciphertexts. In (a), the key is $K_0 = [0.2, 0.3, 0.4, 0.8]$. In (b), the keys are $K_0 = [0.2, 0.3, 0.4, 0.8]$ and $K_1 = [0.2, 0.3 + 10^{-10}, 0.4, 0.8]$. As can be seen, the ciphertext is nearly independent from the plaintext, and the ciphertexts are nearly independent from each other. Because of high sensitivity, each parameter can be more than 16 bits, and the key space is more than 64 bits.

Table 2.1. Randomness test of the produced binary sequence. The sequence is of 20000 bits with the key $K = [0.1, 0.3, 0.6, 0.8]$.

Monobit	Poker	Run test				Long runs
		Run length	Frequency	Run length	Frequency	
9979	20.38	1	4986	7	97	None
		2	2492	8	58	
		3	1279	9	22	
		4	643	10	12	
		5	336	11	3	
		6	179	12	1	

For this cipher, a kind of known-plaintext attack is to solve Eq. 2.1. Taking the i-th plaintext-ciphertext couple for example, in order to obtain ω_i and θ_i, 4 times of attempts should be carried out. Thus, the attacker should attempt for 2^{2N} times in order to obtain P. If N is bigger than 64, the difficulty is no smaller than the brute-force attack, $2^{2N} \geq 2^{128}$. This property keeps the cryptosystem secure against this kind of known-plaintext attack.

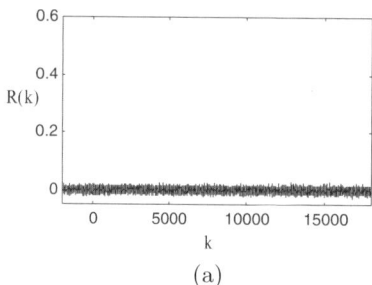 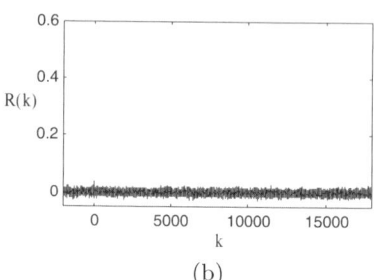

(a)   (b)

Fig. 2.8. Security test of the stream cipher. (a) Correlation between plaintext and ciphertext, (b) Correlation between different ciphertexts

2.3.3 Selective video encryption based on Advanced Video Coding

The proposed encryption scheme

In Advanced Video Coding (AVC) [58], the produced data stream is composed of intra-prediction mode, motion vector difference (MVD), residue data and some ancillary information. Considering that the residue data are often near to zero because of inter/intra-prediction, encrypting only the residue data is not secure enough. Similarly, encrypting only the intra-prediction mode [59] is not secure enough because it is fragile to replacement attacks [60, 61]. To encrypt the intra-prediction mode, MVD and residue data completely may be secure, while it is often time cost. Thus, it is reasonable to encrypt these parameters partially. In the proposed partial encryption scheme, the intra-prediction mode, the residue data's signs and the MVD's signs are encrypted with the proposed chaotic neural network based stream cipher. Here, the intra-prediction modes are encrypted after the entropy encoding, while residue data and MVDs are encrypted after quantization and before entropy encoding.

Security analysis

Compared with the permutation scheme [59], the proposed partial encryption scheme encrypts both the texture information and the motion information, which makes it difficult to recognize the video's content. Fig. 2.9 shows the encryption results of different sample videos, where (b) and (e) are encrypted with the permutation scheme, (c) and (f) are encrypted with the proposed partial encryption scheme. As can be seen, (c) and (f) are more chaotic than (b) and (e), which shows that the proposed scheme is of higher perception security.

The security against replacement attack is analyzed by comparing the proposed scheme and the permutation scheme. For the permutation scheme, only the intra-prediction mode is permuted. By replacing the encrypted prediction

(a) Original   (b) Existing scheme   (c) Proposed scheme

(d) Original   (e) Existing scheme   (f) Proposed scheme

Fig. 2.9. Results of video encryption

modes with certain value, the decoded video Fig. 2.10(a) is intelligible. Differently, besides the prediction mode, the MVD and residue data are encrypted in the proposed scheme, which makes the reconstructed video Fig. 2.10(b) still unintelligible. It proves that the proposed encryption scheme is more secure against replacement attacks.

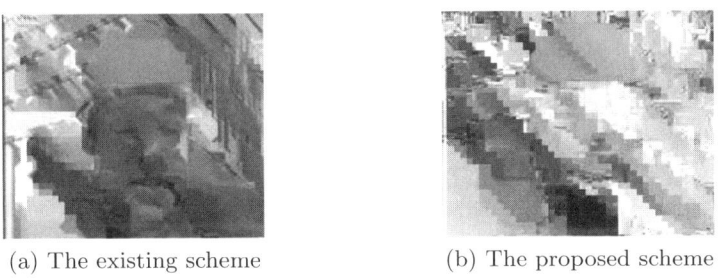

(a) The existing scheme   (b) The proposed scheme

Fig. 2.10. Security against replacement attacks

Time efficiency

The time efficiency of the encryption/decryption process is measured by the time ratio between encryption/decryption and compression/decompression. Seen from Table 2.2, the encryption/decryption operation holds no more than 10% of the compression/decompression operation. It shows that the encryption/decryption process does not affect the compression/decompression process greatly, which makes it practical to combines them together.

Table 2.2. Test of time efficiency

Video	Size	Time ratio	
		Encryption/Compression	Decryption/Decompression
Foreman	QCIF	1.6%	5.4%
Bus	QCIF	2.2%	6.8%
Flower	QCIF	1.2%	6.2%
Akiyo	QCIF	0.9%	5.0%
Mother	QCIF	1.1%	4.7%
Silent	CIF	1.6%	5.7%
Mobile	CIF	1.6%	4.1%
Football	CIF	2.6%	8.2%
Akiyo	CIF	1.9%	6.1%
Foreman	CIF	1.4%	5.3%

2.4 Multimedia content authentication based on chaotic neural networks

2.4.1 Introduction to multimedia content authentication

Different from media encryption that protects media data's confidentiality, media authentication aims to protect media data's integrity. Different from text/binary data authentication, media data (image or video) authentication prefers to content authentication. For example, such operations on an image as recompression or adding noise are often acceptable, while the malicious tempering on an image is not permitted. Thus, media authentication algorithms should be robust to some acceptable operations while sensitive to unacceptable operations. For the special requirement, media authentication technology has been attracting more and more researchers since the past decade.

In image authentication, hash function is often used. Till now, some hash functions have been reported [62–66]. Traditional hash, such as MD5 [66], is sensitive to slight changes. Thus, it is more suitable for the images that permit no modifications. To detect only the tampered regions, the image should be partitioned into regions, and each region produces a hash value with at least 128-bit. The extra channel for these regions' hash is often limited. The hash function proposed by Fridrich et al [64] produces only one bit for each data region, which reduces the length of hash value, but also decreases its security. The one proposed by Lin and Chang [65] uses the traditional hash to authenticate the extracted features between two DCT blocks, which produces at least 128 bits for each DCT block if MD5 is used. This length requires much storage space, and cannot be accepted for some applications. Therefore, contradictions exist between the hash value's length and the security. It is necessary to study new hash that produces suitable length of hash value and also keeps secure. For the one-way property, neural network has been used to construct hash function [67]. In the following content, a hash function based on a chaotic neural network is presented. The hash function can be controlled by the key, and its hash length is also controllable. Thus, the hash function is applied to the features extracted from the image regions, which produces the hash value with a suitable length. The produced hash value is then embedded into the corresponding image region. The scheme can realize self-authentication.

2.4.2 The hash function based on chaotic neural network

The proposed hash function is composed of two parts: data confusion and data diffusion, as shown in Fig. 2.11(a). Here, an $M \times N$ image is partitioned into m data regions, each data region produces a feature vector $P_i(i = 0, 1, \cdots, m - 1)$, and each feature vector is composed of n floating feature pixels: $p_0, p_1, \cdots, p_{n-1}(0 \leq p_i \leq 1)$. Generally, each feature vector is composed of nr bits, and each r bits are transformed into a floating pixel ranging in $[0, 1]$. Then each feature vector P_i is confused firstly with a confusion neuron layer under the control of W_c and B_c. The confusion process generates a string of confused feature vectors $P'_0, P'_1, \cdots, P'_{m-1}$. Then, each of the confused feature vectors is diffused with a diffusion neuron layer under the control of W_d and B_d, which produces a hash value $C'_i(i = 0, 1, \cdots, m-1)$ corresponding to each data region. These regions' hash values are combined to compose a hash vector C that is the produced hash value.

The confusion neuron layer shown in Fig. 2.11(b) is defined as

$$\begin{cases} P'_i = f(W_{ci}P_i + B_{ci}), \\ f(x) = x \end{cases}$$

where '+' denotes module 1 addition, P_i and B_{ci} are composed of floating data ranging in $[0,1]$, and W_{ci} is composed of binary data '0' or '1'. W_{ci} is defined as

$$W_{ci} = \begin{bmatrix} 0 & 1 & 0 & \cdots \\ 0 & \cdots & 1 & 0 \\ \vdots & \cdots \cdots & \vdots \\ 1 & 0 & \cdots & 0 \end{bmatrix}_{n \times n}.$$

Here, there is only one '1' in each row or each column of W_{ci}. Thus, if $B_{ci} = 0$, then W_{ci} permutes P_i only. Otherwise, P_i is both confused and diffused. Here, both W_{ci} and B_{ci} are generated under the control of the key.

Similarly, the diffusion neuron layer shown in Fig. 2.11(c) is defined as

$$\begin{cases} C_i = g(W_{di}P'_i + B_{di}) \\ g(x) = l^q(x) \\ l(x) = 4x(1 - x) \end{cases},$$

where '+' denotes module 1 addition, $l()$ is the chaotic Logistic map with $q \geq 50$, $l^q()$ means to iterate $l()$ for q times, W_{di} and B_{di} are produced by the

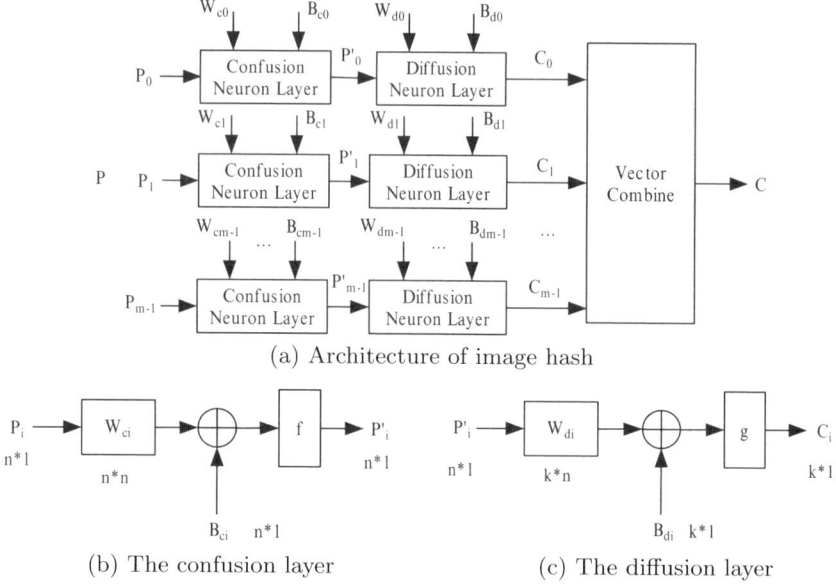

(a) Architecture of image hash

(b) The confusion layer   (c) The diffusion layer

Fig. 2.11. The image hash based on chaotic neural network

key. In the following content of the section, '+' denotes module 1 addition. W_{di} is a matrix of k rows and n columns, which transforms n pixels into k pixels. The final hash value with kr bits is formed by extracting r bits from each of the k pixels. The process is defined as

$$
C_i = \begin{bmatrix} c_{i,0} \\ c_{i,1} \\ \vdots \\ c_{i,k-1} \end{bmatrix} = g(W_{di} P'_i + B_{di}) = g(P''_i)
$$

$$
= g \left(\begin{bmatrix} w_{i,0,0} & w_{i,0,1} & \cdots & w_{i,0,n-1} \\ w_{i,1,0} & w_{i,1,1} & \cdots & w_{i,1,n-1} \\ \vdots & \vdots & \vdots & \vdots \\ w_{i,k-1,0} & w_{i,k-1,1} & \cdots & w_{i,k-1,n-1} \end{bmatrix} \begin{bmatrix} p'_{i,0} \\ p'_{i,1} \\ \vdots \\ p'_{i,n-1} \end{bmatrix} + \begin{bmatrix} b_{i,0} \\ b_{i,1} \\ \vdots \\ b_{i,k-1} \end{bmatrix} \right)
$$

$$
= g \left(\begin{bmatrix} p''_{i,0} \\ p''_{i,1} \\ \vdots \\ p''_{i,k-1} \end{bmatrix} \right).
$$

Thus, in order to decrease the length of image hash, k is often a value smaller than n. And $g()$ is defined as

$$\sum_{j=0}^{k-1} c_{i,j} = g\left(\sum_{j=0}^{k-1} p''_{i,j}\right) = l^q\left(\sum_{j=0}^{k-1} p''_{i,j}\right).$$

For each image, the confusion process and diffusion process are applied to each data region, respectively. Thus, the image hash $H()$ is defined as

$$C = \begin{bmatrix} C_0 \\ C_1 \\ \vdots \\ C_{m-1} \end{bmatrix} = \begin{bmatrix} g(W_{d0}f(W_{c0}P_0 + B_{c0}) + B_{d0}) \\ g(W_{d1}f(W_{c1}P_1 + B_{c1}) + B_{d1}) \\ \vdots \\ g(W_{dm-1}f(W_{cm-1}P_{m-1} + B_{cm-1}) + B_{dm-1}) \end{bmatrix} = H(P, K),$$

where K is the key that produces W_c, B_c, W_d and B_d.

Additionally, the chaotic neural network composed of the confusion neuron and diffusion neuron can be serialized in order to reduce the length of the produced hash value. Thus, if the original image feature is of L_0 bits, the hash value is of L_c bits, and the serial layer is T, then the following condition is satisfied.

$$L_c = L_0 \cdot (k/n)^T$$

Since k is often smaller than n, the image signature with suitable length can be obtained through choosing suitable serial layer T.

2.4.3 The proposed image authentication scheme

Based on the proposed image hash, an image authentication scheme is presented in Fig. 2.12. At the sender end, the feature P is extracted from an image V, from which, the hash value C is computed under the control of K_1. Then, C is embedded into the image V under the control of K_2. The process is described as

$$S = Emb(V, C, K_2) = Emb(V, H(P, K_1), K_2) = Emb(V, H(Fea(V), K_1), K_2),$$

where S is the embedded image, $Fea()$ is the feature extraction process, and $Emb()$ is the embedding algorithm. The extracted feature determines the robustness of the authentication process. Taking wavelet transformation for example, in order to keep robust to recompression and sensitive to the malicious tampering, the high bit-planes of the code blocks in the low frequency-band are preferred. For example, if an image is transformed by 4-level wavelet transformation, then only the 7 highest bit-planes of the code blocks in the 10 lowest frequency-bands are selected as features. The computed hash value should be embedded into the image itself imperceptibly [68], which saves extra space for hash value's storage. To keep imperceptibility, the signature is embedded into the position with little effect on images' quality, for example, the 8-th highest bit-plane. The embedding position is controlled by the key.

Similarly, at the receiver end, the embedded hash value C is extracted from the image, which is compared with the computed one C'. If they are same to

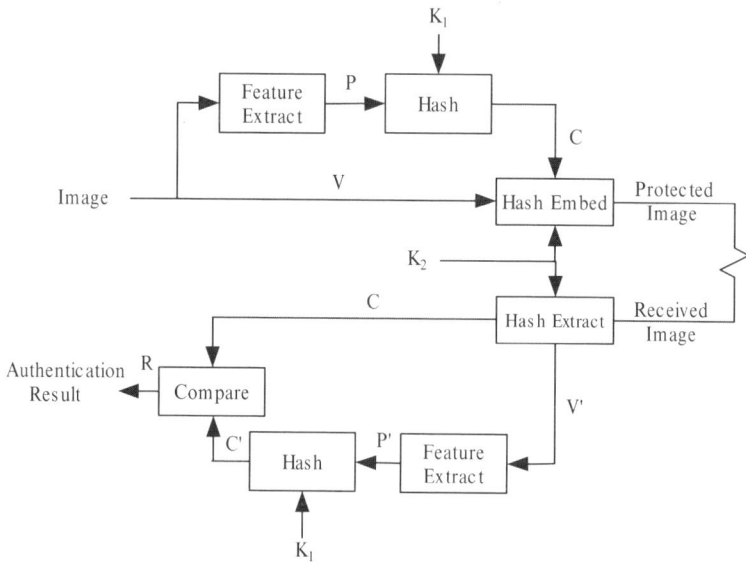

Fig. 2.12. The proposed image authentication scheme

each other, then the received image is authenticated. Otherwise, the image's integrity is destroyed, and the difference between C and C' determines the regions that have been tampered. The authentication process is defined as

$$\begin{cases} (C, V') = Ext(S, K_2) \\ R = Com(C, C') = Com(C, H(P', K_1)), \end{cases}$$

where R is the authentication result, $Ext()$ is the extraction algorithm [68], and $Com()$ is to determine the difference between C and C'.

2.4.4 Performance analysis

Security

In the proposed hash function, the confusion layer and diffusion layer are combined together to obtain high computing security, which obeys the cryptographic design principles [57]. It is different from the function proposed in [67], which depends on the randomness of the sequence produced by a chaotic neural network. Some other neural networks with more confusion layers or diffusion layers can also be used to design a hash function with higher security.

The proposed image hash is nearly a one-way function. That is, it is difficult to recover V from C. The signature process is controlled by the K, which makes brute-force attack difficult. For attackers knowing both P and C, the

difficulty to solve a singular equation (the diffusion function) makes the attacks unpractical. For select-image attackers, they have to break the confusion
process and to decode P'' from C before to solve the singular equation. But
the brute-force space of confusion process is $n \cdot (n-1) \cdot (n-2) \cdot \ldots \cdot 1$, which
is bigger than 2^n if $n > 4$. In practice, the feature vector is often of more
than 16 pixels, which makes brute-force attacks difficult. What's more, the iterated Logistic map is difficult to solve, which requires 2^q times of brute-force
attempts. Thus, the brute-force space of decoding P'' is no smaller than 2^{50}
when $q \geq 50$. And the overall space of select-image attack is 2^{n+q} that make
it difficult to falsify an image hash.

The hash function is of high plaintext sensitivity. The confusion process
and diffusion process make the signature sensitive to the plaintext. From the
diffusion function, we get

$$c_{i,j} = g(\sum_{t=0}^{n-1} w_{i,j,t} p'_{i,t} + b_{i,j}),$$

where j varies from 0 to k-1. As can be seen, $c_{i,j}$ is in relation with $p'_{i,t} (t =
0, 1, \cdots, n-1)$, that is, a bit of difference in the plaintext causes great dif-
ferences in the hash value. Additionally, in the proposed hash function, the
control parameters are independent of the plaintext. These properties make
attackers difficult to find another plaintext with the same signature.

The image hash is controlled by the key. That is, for the same image,
various hash values can be generated, while only the user with the correct
key can authenticated it. This property makes it difficult for attackers to
falsify hash values. The same advantage can be drawn from the key-based
hash embedding process. Only the authenticated user can embed or extract
the hash value correctly.

Authentication performance

The proposed image hash can keep robust to acceptable operations, such as recompression or slight noise. Table 2.3 gives the experimental results on various
images, where the embedding bit-plane varies from 6 to 8 and (n,k,r)=(64,4,4).
The 10 test images include Lena, Airplane, Couple, Girl, Boats, Cameraman,
and so on. The hash length of each data region is 16. Here, the JPEG2000
compression ratio changes from 5 to 9, and the zero-mean Gaussian noise's
variance varies from 0.1 to 0.0001. As can be seen, when 6 or 7 bit-planes are
used as the feature, all the test images are correctly authenticated. Only 7
or 5 images are correctly authenticated when 8 bit-planes are hashed and the
compression ratio is no smaller than 8. In practice, $C \leq 8$ and $N \leq 0.01$ are
acceptable operations, and now the number of the hashed bit-planes should
be no bigger than 7.

Additionally, this scheme is sensitive to malicious tampering, such as cutting, pasting or substitution. Fig. 2.13 gives the experimental results on image

Table 2.3. Robustness of the proposed image hash

Compress, Noise	6 Bit-planes		7 Bit-planes		8 Bit-planes	
	Compress	Noise	Compress	Noise	Compress	Noise
5, 0.0001	10	10	10	10	10	10
6, 0.001	10	10	10	10	10	10
8, 0.01	10	10	10	10	7	7
9, 0.1	10	10	10	10	5	4

Plane, where the image is tampered by cutting and substitution, the number of the hashed bit-planes varies from 6 to 8, and $(n,k,r)=(64,4,4)$. As can be seen, two tampered data regions are not detected if only 6 bit-planes are hashed, while all the tampered regions are detected when 7 or 8 bit-planes are hashed. It shows that the more the hashed bit-planes, the higher the system's sensitivity. Through experiments on various images, the number of the hashed bit-planes is recommended as 7.

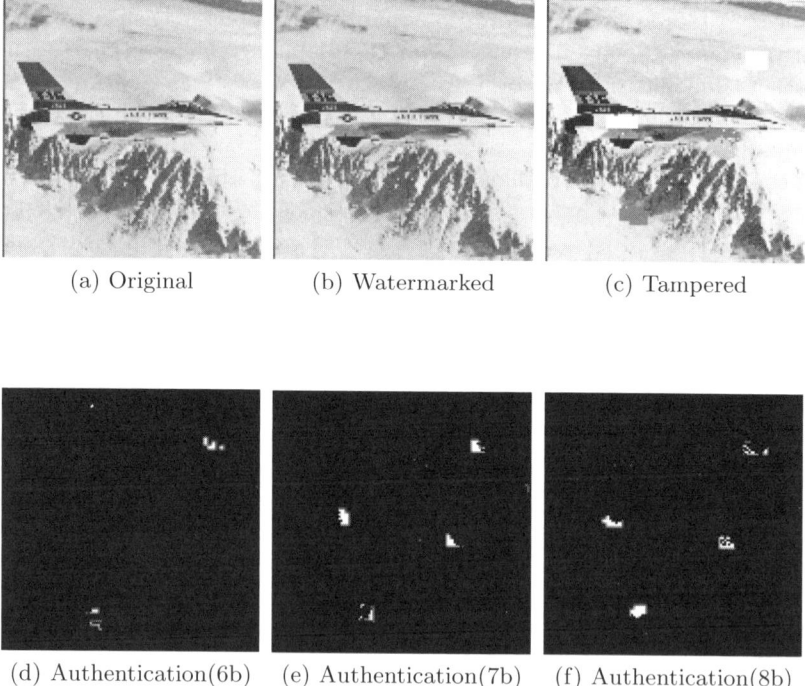

(a) Original   (b) Watermarked   (c) Tampered

(d) Authentication(6b)   (e) Authentication(7b)   (f) Authentication(8b)

Fig. 2.13. Sensitivity of the authentication scheme. (a) original, (b) embedded, (c) tampered, (d) authentication result of 6 bit-planes, (e) authentication result of 7 bit-planes, and (f) authentication result of 8 bit-planes

Computing complexity

The computing complexity depends on the data volumes to be hashed. Taking the time ratio between image authentication and image compression as a measurement, the results are shown in Table 2.4. Here, the embedding bit-plane varies from 6 to 8, (n,k,r)=(64,4,4), and JPEG2000 is applied under the compression ratio of 8. As can be seen, the time ratio increases with the rise of the number of hashed bit-planes, and it is no bigger than 20% when the number of the hashed bit-plane is not bigger than 7. Thus, the authentication process does not affect compression process greatly, which makes it practical to combine them together to realize real time image compression and authentication.

Table 2.4. Test of computing complexity

Image(Size,Color)	Time ratio of authentication/compression		
	6 Bit-planes	7 Bit-planes	8 Bit-planes
Lena (256×256, Colorful)	9.2%	11.6%	18.6%
Airplane (256×256, Colorful)	10.8%	16.4%	21.9%
Couple (256×256, Colorful)	8.9%	12.7%	16.8%
Girl (128×128, Gray)	9.4%	16.3%	19.9%
Boats (512×512, Gray)	10.5%	15.6%	22.8%
Cameraman (256×256, Gray)	7.8%	14.7%	18.5%

2.5 Future work and discussions

Considering that multimedia data need special encryption or authentication algorithms that are time efficient and support the plaintext with variable size, some new data encryption or authentication algorithms are expected. In this chapter, the cipher and hash function based on chaotic neural networks are investigated, and proved workable. However, for further practicability, some improvements are expected.

Firstly, the implementation in integer form is preferred, which is easier to realize digital information protection. Now, most of the chaotic neural network based algorithm is implemented in floating form, which makes the system rely on the word length of the computer. In integer form, the computing resolution is confirmed, and it is easy to realize synchronization between different computers. The difficulty is to make the integer implementation keep the same performance with the floating implementation.

Secondly, the parallel property is to be used. Compared with traditional ciphers, chaotic neural network's parallel implementation makes the ciphers based on chaotic neural networks easy in efficiency improvement. To adopt the parallel structure to realize lightweight encryption is an attractive and challenging work.

Thirdly, the chaotic neural networks with more input neurons and neural layers are preferred. In the chapter, only some simple chaotic neural networks are considered. In fact, the neural networks with more complicated structure, such as more layers or more input neurons, may bring higher security. Of course, the computing efficiency should be considered in order to obtain suitable tradeoff.

Fourthly, the security of the ciphers based on chaotic neural networks should be analyzed thoroughly. Besides such traditional analysis methods for ciphers as brute-force space, linear analysis or differential analysis, some special attacks should be considered, such as chaos mode analysis [69] or neural network structure analysis. This work should be done to strengthen the ciphers based on chaotic neural networks.

Fifthly, the partial encryption scheme should be carefully designed. According to partial encryption, the encrypted media data should be of high perception security, and keeps independent from the unencrypted data. Additionally, such media data with fixed pattern are not suitable for encryption, such as the file header or synchronization information, because they are fragile to replacement attacks. Thus, the encryption scheme should be considered besides that the cipher is secure.

2.6 Conclusions

In this chapter, the chaotic neural network's application in media encryption and authentication is investigated. By using the chaotic neural network's properties, such as diffusion, confusion, one-way, parameter sensitivity, random similarity, the cipher and hash function suitable for media encryption and authentication are designed. Based on the cipher, the partial encryption scheme for AVC video is proposed and analyzed. The experimental results show the scheme is secure in perception and efficient in practice. Based on the hash function, the image authentication scheme is proposed to authenticate media data's integrity. The analyses and experiments show that the scheme is secure against malicious tampering and robust to such acceptable operations as compression or noise. These attempts prove the practicability of media encryption based on chaotic neural networks.

Although chaotic neural network's properties show its potential value in media data encryption, using chaotic neural network to encrypt media data is just beginning, with much work to be done. The efficiency and computing resolution are to be improved by changing the floating implementation into integer implementation. Additionally, the security should be thoroughly analyzed by considering of both traditional attacks and some special attacks.

However, considering of the urgent need of media protection, it should be confirmed that the future of the research work is still bright.

2.7 Acknowledgements

The author wants to thank Prof. Z. Wang and Mr. J. Wang for their great help and to thank reviewers for their valuable advices.

References

1. Babloyantz A, Salazar J, Nicolis C (1985) Evidence for Chaotic Dynamics of Brain Activity during the Sleep Cycle. Phys Lett 111:152–156
2. Wasserman P (1993) Advanced Methods in Neural Computing. New York: Van Nostrand Reinhold, New York
3. Chay T, Fan Y (1993) Evolution of Periodic States and Chaos in Two Types of Neural Models. In: Ditto W(ed.) Chaos in Biology and Medicine, Proc SPIE, Washington: SPIE, 2036:100–114
4. Aibara K (1990) Chaotic Neural Networks. Physical Letter A 144(6,7):334–340.
5. Inoue M, Nagayoshi A (1991) A Chaos Neuro-computer. Physical Letter A 158(8):373–376
6. Inoue M, Nakamoto K (1993) Epilepsy in a Chaos Neuro-computer Model. In SPIE Proceeding, Chaos in Biology and Medicine, 2306:77–84
7. Chan C, Cheng L (1998) Pseudorandom Generator Based on Clipped Hopfield Neural Network. In: Proceedings of the 1998 IEEE International Symposium on Circuits and Systems, 3:183–186
8. Chan C, Chan C, Cheng L (2001) Software Generation of Random Number by using Neural Network. In Proceeding of 5th International Conference on Artificial Neural Networks and Genetic Algorithms (ICANNGA), Pragus, Czech Republic
9. Guo D, Cheng L, Cheng L (1999) A New Symmetric Probabilistic Encryption Scheme Based on Chaotic Attractors of Neural networks. Applied Intelligence 10(1):71–84
10. Xiang T, Liao X, Tang G, Chen Y, Wong K (2006) A novel block cryptosystem based on iterating a chaotic map. Physics Letters A 349(1-4):109–115
11. Xiao D, Liao X, Deng S (2005) One-way Hash function construction based on the chaotic map with changeable-parameter. Chaos, Solitons & Fractals 24(1):65–71
12. Rachel M, Einat K, Wolfgang K (2003) Public Channel Cryptography by Synchronization of Neural Networks and Chaotic Maps. Physical Review Letters 91(11):118701/1–118701/4
13. Cauwenberghs G (1999) Delta-sigma cellular automata for analog VLSI random vector generation. IEEE Transactions on Circuits and Systems II: Analog and Digital Signal Processing 46(3):240–250
14. Chan C, Cheng L (2001) The convergence properties of a clipped Hopfield network and its application in the design of keystream generator. IEEE Transactions on Neural Networks 12(2):340–348

15. Karras D, Zorkadis V (2003) On neural network techniques in the secure management of communication systems through improving and quality assessing pseudorandom stream generators. Neural Networks 16(5-6):899–905
16. Caponetto R, Lavorgna M, Occhipinti L (1996) Cellular neural networks in secure transmission applications. In Proceedings of the IEEE International Workshop on Cellular Neural Networks and their Applications, 411–416
17. Yue T, Chiang S (2000) A neural network approach for visual cryptography. In Proceedings of the IEEE-INNS-ENNS International Joint Conference on Neural Networks 5:494–499
18. Yen J, Guo J (1999) A Chaotic Neural Network for Signal Encryption/Decryption and Its VLSI Architecture. In: Proc. 10th (Taiwan) VLSI Design/CAD Symposium, 319-322
19. Lian S, Chen G, Cheung A, Wang Z (2004) A Chaotic-Neural-Network-Based Encryption Algorithm for JPEG2000 Encoded Images. In Proceedings of 2004 IEEE Symposium on Neural Networks (ISNN2004),Springer LNCS, 3174:627–632
20. Yee L, Silva D (2002) Application of multilayer perception networks in symmetric block ciphers. In Proceedings of the 2002 International Joint Conference on Neural Networks, 2:1455–1458
21. Vanstone S, Menezes A, Oorschot P (1996) Handbook of Applied Cryptography. CRC Press
22. Shannon C (1949) Communication theory of secrecy systems. Bell System Technical Journal 28:656–715
23. Li C, Li S, Zhang D, Chen G (2004) Cryptanalysis of a Chaotic Neural Network Based Multimedia Encryption Scheme. In Proceedings of the 2004 Pacific-Rim Conference on Multimedia (PCM2004), Springer LNCS, 3333:418–425
24. Fridrich J (1998) Symmetric ciphers based on two-dimensional chaotic maps. Int. J. Bifurcation and Chaos 8(6):1259–1284
25. Chen G, Mao Y, Chui C (2004) A symmetric image encryption scheme based on 3D chaotic cat maps. Chaos, Solitons and Fractals 12:749–761
26. Li S, Zheng X, Mou X, Cai Y (2002) Chaotic Encryption Scheme for Real-time Digital Video. In Proceedings of SPIE, Electronic Imaging, Real-time Imaging V1, 4666:149–160
27. Ren X, Hu G, Tan Z (2000) Synchronization of chaotic neural network and its application in secure communication. Journal of Shanghai Jiaotong University 34(6):744–747
28. Elman J (1990) Finding structure in time. Cognitive Science 14:179-211
29. Pyragas K (1992)Continuous control of chaos by self controlling feedback. Phys Lett A 170:421–428
30. Ushio T (1995) Chaotic synchronization and controlling chaos based on contraction maps. Phys Lett A 198:14–22
31. Kocarev L, Parlitz U (1995) General approach for chaotic synchronization with application to communication. Phys Rev Lett 74(25):5028–5031
32. Zou F, Nossek J (1991) A Chaotic Attractor with Cellular Neural Networks. IEEE Transaction on Circuits and Systems 38(7):811–812
33. Zou F, Joset A (1993) Nossek. Bifurcation and chaos in Cellular Neural Networks. IEEE Transaction on Circuits and Systems-I: Fundamental Theory and Applications 40(3):166–173
34. Kaneko K (1990) Clustering, coding, switching, hierarchical ordering and control in a network of chaotic elements. Physics D 41(22):137–172

35. Aihara K, Takabe T, Toyoda M (1990) Chaotic Neural Networks. Phys Lett A 144(6):333–340
36. Inoue M, Nagayoshi A (1991) A Chaos Neuro-computer. Phys Lett A 158(8):373–376
37. Addison P (1997) Fractals and Chaos - an illustrated course, Institute of physics publishing
38. Fang Y, Yu Y (1993) Dynamic analysis of chaotic neurons. In Proceedings of Chinese Conference on Neural Networks, 199-203
39. Nagayama I (1995) Spectral Analysis of the Chaotic Behavior Obtained by Neural Networks. In Proceedings of 1995 International Conference on Neural Networks, 28–31
40. Kudelski A (1994) Method for scrambling and unscrambling a video signal. United States Patent 5375168
41. Access control system for the MAC/packet family: EUROCRYPT. European Standard EN 50094, CENELEC (1992)
42. Maniccam S, Bourbakis N (2004) Image and video encryption using SCAN patterns. Pattern Recognition 37(4):725–737
43. Qiao L, Nahrstedt K (1997) A new algorithm for MPEG video encryption. In Proceeding of the First International Conference on Imaging Science, Systems and Technology (CISST'97),21–29
44. Romeo R, Romdotti G, Mattavelli M, Mlynek D (1999) Cryptosystem architectures for very high throughput multimedia encryption: the RPK solution. In Proceedings of ICECS'99, 1:261–264
45. Shi C, Wang S, Bhargava, B (1999) MPEG video encryption in real-time using secret key cryptography. In Proc. of International Conference of Parallel and Distributed Processing Techniques and Applications (PDPTA'99), 2822–2828
46. Chiaraluce F, Ciccarelli L, Gambi E, Pierleoni P, Reginelli M (2002) A New Chaotic Algorithm for Video Encryption. IEEE Transactions on Consumer Electronics 48(4):838–844
47. Yen J, Guo J (1999) A new MPEG encryption system and its VLSI architecture. In Proceeding of IEEE Workshop on Signal Processing Systems,430–437
48. Tosum A, Feng W (2000) Efficient multi-layer coding and encryption of MPEG video streams. In Proceedings of IEEE International Conference on Multimedia and Expo (I), 119–122
49. Qiao L, Nahrstedt K, Tam I (1997) Is MPEG encryption by using random list instead of zigzag order secure. In Proceeding of IEEE International Symposium on Consumer Electronics, 226–229
50. Shi C, Bhargava B (1998) A fast MPEG video encryption algorithm. In Proceedings of the 6th ACM International Multimedia Conference, 81–88
51. Zeng W, Lei S (2003) Efficient frequency domain selective scrambling of digital video. IEEE Trans Multimedia 5(1):118–129
52. Wu C, Kuo C (2000) Fast encryption methods for audiovisual data confidentiality. In SPIE International Symposia on Information Technologies 2000, Proceedings of SPIE 4209:284–295
53. Wu C, Kuo C (2001) Efficient multimedia encryption via entropy codec design. In SPIE International Symposium on Electronic Imaging 2001, Proceedings of SPIE 4314
54. Lian S, Sun J, Wang Z (2004) Perceptual Cryptography on SPIHT Compressed Images or Videos. In Proceedings of 2004 IEEE Conference on Multimedia & Expo (ICME2004), 3:2195–2198

55. Lian S, Sun J, Wang Z (2004) A Novel Image Encryption Scheme Based-on JPEG Encoding. In Proceedings of the Eighth International Conference on Information Visualization (IV), 217–220
56. Li S, Chen G, Mou X (2005) On the Dynamical Degradation of Digital Piecewise Linear Chaotic Maps. International Journal of Bifurcation and Chaos 15(10):3119-3151
57. Security Requirements for Cryptographic Modules (Change Notice). Federal Information Processing Standards Publication (FIPS PUB) 140-1 (2001)
58. ITU-T Rec. H.264/ISO/IEC 11496-10. Advanced Video Coding. Final Committee Draft, Document JVT-E022 (2002)
59. Ahn J, Shim H, Jeon B, Choi I (2004) Digital Video Scrambling Method Using Intra Prediction Mode. PCM2004, Springer, LNCS 3333:386–393
60. Lian S, Sun J, Zhang D, Wang Z (2004) A Selective Image Encryption Scheme Based on JPEG2000 Codec. In Proceedings of 2004 Pacific-Rim Conference on Multimedia, Springer LNCS 3332:65–72
61. Podesser M, Schmidt H, Uhl A (2002) Selective Bitplane Encryption for Secure Transmission of Image Data in Mobile Environments. In: CD-ROM Proceedings of the 5th IEEE Nordic Signal Processing Symposium (NORSIG 2002). Tromso-Trondheim, Norway
62. Fridrich J, Goljan M, Du R (2001) Invertible Authentication. In: Proceeding of SPIE, Security and Watermarking of Multimedia Contents, San Jose, California, 3971:197–208
63. Fridrich J (2000) Visual Hash Functions for Obvious Watermarking. In: Proceeding of SPIE Photonic West Electronic Imaging 2000, Security and Watermarking of Multimedia Contents, San Jose, California, 286–294
64. Fridrich J, Goljan M (2000) Robust Hash Functions for Digital Watermarking. The International Conference on Information Technology: Coding and Computing Las Vegas, Nevada, 178–183
65. Lin C, Chang S (1998) A Robust Image Authentication Method Surviving JPEG Lossy Compression. In: Proceedings of SPIE Storage and Retrieval of Image/Video Databases, San Jose, 3312:296–307
66. Rivest R (1992) The MD5 Message Digest Algorithm. RFC 1321
67. Xiao D, Liao X (2004) A Combined Hash and Encryption Scheme by Chaotic Neural Network. In Proceedings of ISNN2004, 2:633–638
68. Yeung M, Mintzer F (1997) An Invisible Watermarking Technique for Image Verification. In: Proceeding of ICIP'97, Santa Barbara, California, 2:680–683
69. Zhusubaliyev Z, Mosekilde E (2003) Bifurcations and Chaos in Piecewise-Smooth Dynamical Systems. World Scientific, Singapore.

Evolutionary Regular Substitution Boxes

Nadia Nedjah[1] and Luiza de Macedo Mourelle[2]

[1] Department of Electronics Engineering and Telecommunications,
Engineering Faculty,
State University of Rio de Janeiro,
Rua São Francisco Xavier, 524, Sala 5022-D,
Maracanã, Rio de Janeiro, Brazil
nadia@eng.uerj.br, http://www.detel.eng.uerj.br
[2] Department of System Engineering and Computation,
Engineering Faculty,
State University of Rio de Janeiro,
Rua São Francisco Xavier, 524, Sala 5022-D,
Maracanã, Rio de Janeiro, Brazil
ldmm@eng.uerj.br, http://www.desc.eng.uerj.br

A substitution box (or S-box) is simply a transformation of an input sequence of bits into another. The input and output sequences have not necessarily the same number of bits. In cryptography, an S-box constitutes a cornerstone component of symmetric key algorithms. In block ciphers, they are typically used to obscure the relationship between the plaintext and the ciphertext. Non-linear and non-correlated S-boxes are the most secure linear and differential cryptanalysis. In this chapter, we focus on engineering regular S-boxes, presenting high non-linearity and low auto-correlation properties using evolutionary computation. Hence, there are three properties that need to be optimised: regularity, non-linearity and auto-correlation. We use the Nash equilibrium-based multi-objective evolutionary algorithm to engineer resilient substitution boxes.

3.1 Introduction

In cryptography, confusion and diffusion are two important properties of a secure cipher as identified in [1]. Confusion allows one to make the relationship between the encryption key and ciphertext as complex as possible while diffusion allows one to reduce as much as possible the dependency between the plaintext and the corresponding ciphertext. Substitution (a plaintext symbol is replaced by another) has been identified as a mechanism for primarily

N. Nedjah and L. de Macedo Mourelle: *Evolutionary Regular Substitution Boxes*, Studies in Computational Intelligence (SCI) **57**, 79–88 (2007)
www.springerlink.com © Springer-Verlag Berlin Heidelberg 2007

confusion. Conversely transposition (rearranging the order of symbols) is a technique for diffusion. In modern cryptography, other mechanisms are used, such as linear transformations [1]. Product ciphers use alternating substitution and transposition phases to achieve both confusion and diffusion respectively. Here we concentrate on confusion using non-linear and non-correlated substitution boxes or simply S-boxes.

It is well-known that the more linear and the less auto-correlated the S-box is, the more resilient the cryptosystem that uses them. However, engineering a regular S-box that has the highest non-linearity and lowest auto-correlation properties is an *NP*-complete problem. Evolutionary computation is the ideal tool to deal with this type of problems. As there are three objectives that need to be reached, which are maximal regularity, maximal non-linearity yet minimal auto-correlation, we propose to use multi-objective evolutionary optimisation. Therefore, we exploit the game theroy [3] and more specifically the well-known Nash equilibrium strategy [4] to to engineer such *resilient* substitution boxes.

The rest of this chapter is organised in five sections. First, in Section 3.2, we define more formally S-boxes as well as their characteristics. Subsequently, in Section 3.3, we present the multi-objective evolutionary algorithm used to perform the evolution i.e., the Nash equilibrium-based evolutionary algorithm [3]. Thereafter, in Section 3.4, we describe the S-box encoding and the genetic operators used followed by the definition and implementation of an S-box fitness evaluation with respect to all three considered objectives. Then, in Section 3.5, we evaluate the performance of the evolutionary process and assess the quality of the evolved resilient S-boxes. Also, we compare the characteristics of the engineered S-boxes to those of Advanced Encryption Standard (AES) [1] and Data Encryption Standard (DES) [2]. Last but not least, in Section 3.6, we summarise the content of the chapter and draw some useful conclusions.

3.2 Preliminaries for Substitution Boxes

S-Boxes play a basic and fundamental role in many modern block ciphers. In block ciphers, they are typically used to obscure the relationship between the plaintext and the ciphertext. Perhaps the most notorious S-boxes are those used in DES [2]. S-boxes are also used in modern cryptosystems based on AES [1] and Kasumi [5]. All three are called Feistel cryptographic algorithms [6] and have the simplified structure depicted in Fig. 3.1.

An S-box can simply be seen as a Boolean function of n inputs and m outputs, often with $n > m$. Considerable research effort has been invested in designing resilient S-boxes that can resist the continuous cryptanalyst's attacks. In order to resist linear and differential cryptanalysis [7, 8], S-boxes need to be confusing or non-linear and diffusing or non auto-correlated.

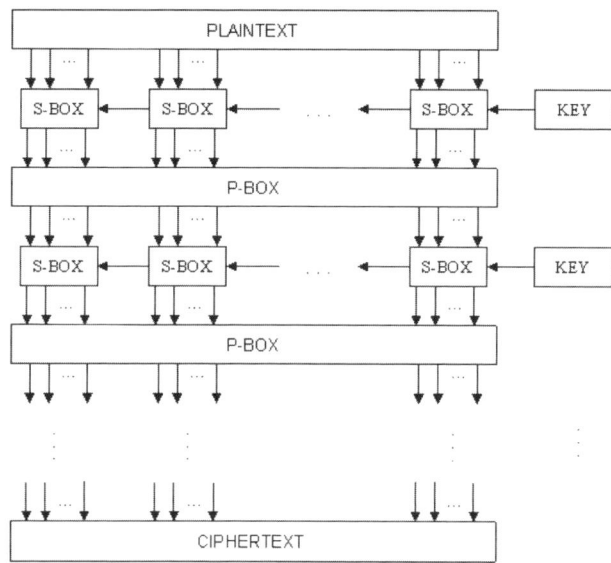

Fig. 3.1. The simplified structure of Feistel cryptographic algorithm

S-boxes also need to be non-regular.

Definition 1. A *simple* S-box $\mathcal{S}$ is a Boolean function defined as $\mathcal{S} \colon \mathcal{B}^n \longmapsto \mathcal{B}$.

Definition 2. A *linear* simple S-boxes $\mathcal{L}$ is defined in (3.1):

$$\mathcal{L}_\beta(x) = \bigoplus_{i=0}^{i=m} \beta_i.\mathcal{L}(x_i) \tag{3.1}$$

Definition 3. The *polarity* of a simple S-box $\mathcal{S}$ is defined in (3.2):

$$\hat{\mathcal{S}}(x) = (-1)^{\mathcal{S}(x)} \tag{3.2}$$

Definition 4. The *uncorrelation factor* of Two simple S-boxes $\mathcal{S}$ and $\mathcal{S}'$ is defined in (3.3):

$$\mathcal{U}_{\mathcal{S},\mathcal{S}'} = \sum_{x \in \mathcal{B}^n} \hat{\mathcal{S}}(x) \times \hat{\mathcal{S}}'(x) \tag{3.3}$$

Definition 5. Two simple S-boxes $\mathcal{S}$ and $\mathcal{S}\prime$ are said to be uncorrelated if and only $\mathcal{U}_{\mathcal{S},\mathcal{S}'} = 0$.

Definition 6. The *non-linearity* of a simple S-box $\mathcal{S}$ is measured by its un-correlation factor with all possible linear simple S-Boxes defined in (3.4):

$$\mathcal{N}_S = \frac{1}{2}\left(2^n - \max_{\alpha \in \mathcal{B}^n} |\mathcal{U}_{S,\mathcal{L}}|\right) \tag{3.4}$$

Definition 7. The *auto-correlation* of a simple S-box S is measured by its uncorrelation factor with derivatives S-boxes $\mathcal{D}(x) = S(x) \oplus \alpha$, for all $\alpha \in \mathcal{B}^n \setminus \{0^n\}$ and defined in (3.5):

$$\mathcal{A}_S = \max_{\alpha \in \mathcal{B}^n \setminus \{0^n\}} |\mathcal{U}_{S,\mathcal{D}}| \tag{3.5}$$

Note that $\mathcal{U}_{S,\mathcal{D}}$ is also called *Walsh Hadamard transform* [9].

Definition 8. A simple S-box is said to be *balanced* if and only the number of combinations $x \in \mathcal{B}^n$ such that $S(x) = 0$ and the number of combinations $y \in \mathcal{B}^n$ such that $S(y) = 1$ are the same. The *balance* of a simple S-box is measured using its *Hamming weight*, defined in (3.6).

$$\mathcal{W}_S = \frac{1}{2}\left(2^n - \sum_{x \in \mathcal{B}^n} \hat{S}(x)\right) \tag{3.6}$$

3.3 Nash Equilibrium-based Evolutionary Algorithms

This approach is inspired by the Nash strategy for economics and game theory [3,4]. The multi- objective optimisation process based on this strategy is non-cooperative in the sense that each objective is optimised separately. The basic idea consists of associating an agent or player to every objective. Each agent attempts to optimise the corresponding objective fixing the other objectives to their best values so far. As proven by Nash in [3], the Nash equilibrium point should be reached when no player can improve further the corresponding objective.

Let m be the number of objectives $f_1, \ldots, f_m$. The multi-objective genetic algorithms based on Nash strategy assigns the optimisation of objective f_i to $player_i$, each of which has its own population. The process is depicted in Figure 3.2. Basically, it is a parallel genetic algorithm [10] with the exception that there are several criteria to be optimised. When a player, say $player_i$, completes an evolution generation, say t, it sends the local best solution reached f_i^t to the all $player_j$, $j \in \{1, \ldots, m\} \setminus \{i\}$, which will then be used to fix objective f_i to f_i^t during the next generation $t + 1$. This evolutionary process is repeated iteratively until no player can further improve the associated criteria.

3.4 Evolving Resilient S-Boxes

In general, two main important concepts are crucial to any evolutionary computation: individual encoding and fitness evaluation. One needs to know how

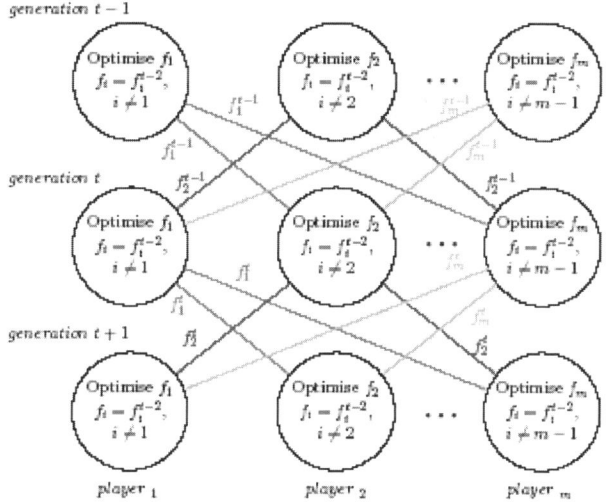

Fig. 3.2. Multi-objective optimisation using Nash strategy

to appreciate the solutions with respect to each one of the multiple objectives. So In the section, we concentrate on these two aspects for evolving optimal S-boxes.

3.4.1 S-Box encoding and genetic operators

We encode an S-box simply using a matrix of bytes. The mutation operator chooses an entry randomly and changes its value, using a randomised byte. Crossover of S-Boxes is implemented using four-point crossover. This is described in Fig. 3.3. The four-point crossover can degenerate to either the triple, double or single-point crossover. Moreover, these can be either *horizontal* or *vertical*.

The four-point crossover degenerates into a triple-point one whenever one of the crossover points (say p) coincides with one of the sides of the S-box. It is a horizontal triple-point when p is one of the vertical crossover points, as in Fig. 3.4–(a) and Fig. 3.4–(b) and horizontal otherwise, as in Fig. 3.4–(c) and Fig. 3.4–(d).

The four-point crossover degenerates into a double-point one whenever either both horizontal points or vertical points coincide with the low and high or left and right limits of the S-box respectively. The horizontal double-point crossover is described in Fig. 3.5–(a) and the horizontal one in Fig. 3.5–(b).

The four-point crossover can also yield a horizontal and vertical single-point crossover as described in Fig. 3.6–(a) and Fig. 3.6–(b) respectively. The vertical single-point crossover happens when the horizontal crossover points

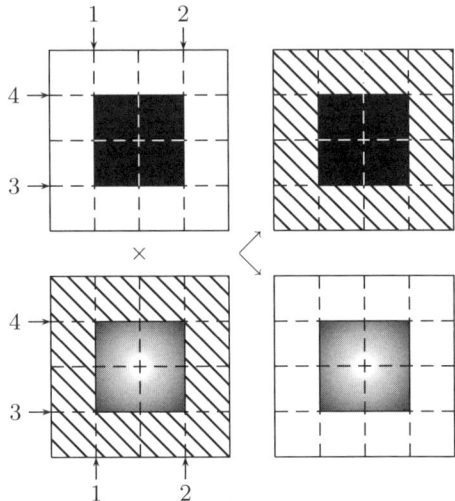

Fig. 3.3. Four-point crossover of S-boxes

coincide with the low and high limits of the S-box and the one of the vertical points coincides with the left or right limit of the S-box. On the other hand, the horizontal single-point crossover happens when the vertical crossover points coincide with the left and right limits of the S-box and one of the horizontal crossover points coincide with low or high limit of the S-box.

3.4.2 S-Box evaluation

Now, let us generalise the definitions of balance, non-linearity and auto-correlation to non-simple S-boxes, i.e. S-boxes defined as $\mathcal{S} \colon \mathcal{B}^n \longmapsto \mathcal{B}^m$.

Definition 9. An S-box $\mathcal{S}$ defined as $\mathcal{S} \colon \mathcal{B}^n \longmapsto \mathcal{B}^m$ is a concatenation of m simple S-boxes $\mathcal{S}_i$ with $1 \leq i \leq m$, such as in (3.7):

$$\mathcal{S}(x) = \mathcal{S}_1(x)\mathcal{S}_2(x)\ldots\mathcal{S}_m(x) \tag{3.7}$$

Definition 10. The *non-linearity* of S-box $\mathcal{S}$ is measured by $\mathcal{N}_{\mathcal{S}}^*$ defined in (3.8):

$$\mathcal{N}_{\mathcal{S}}^* = \min_{\beta \in \mathcal{B}^m \setminus \{0^m\}} \mathcal{N}_{\mathcal{S}_\beta(x)}, \text{ wherein}$$
$$\mathcal{S}_\beta(x) = \bigoplus_{i=0}^{i=m} \beta_i \mathcal{S}_i(x) \tag{3.8}$$

Definition 11. The *auto-correlation* of S-box $\mathcal{S}$ is measured by $\mathcal{A}_{\mathcal{S}}^*$ defined in (3.9):

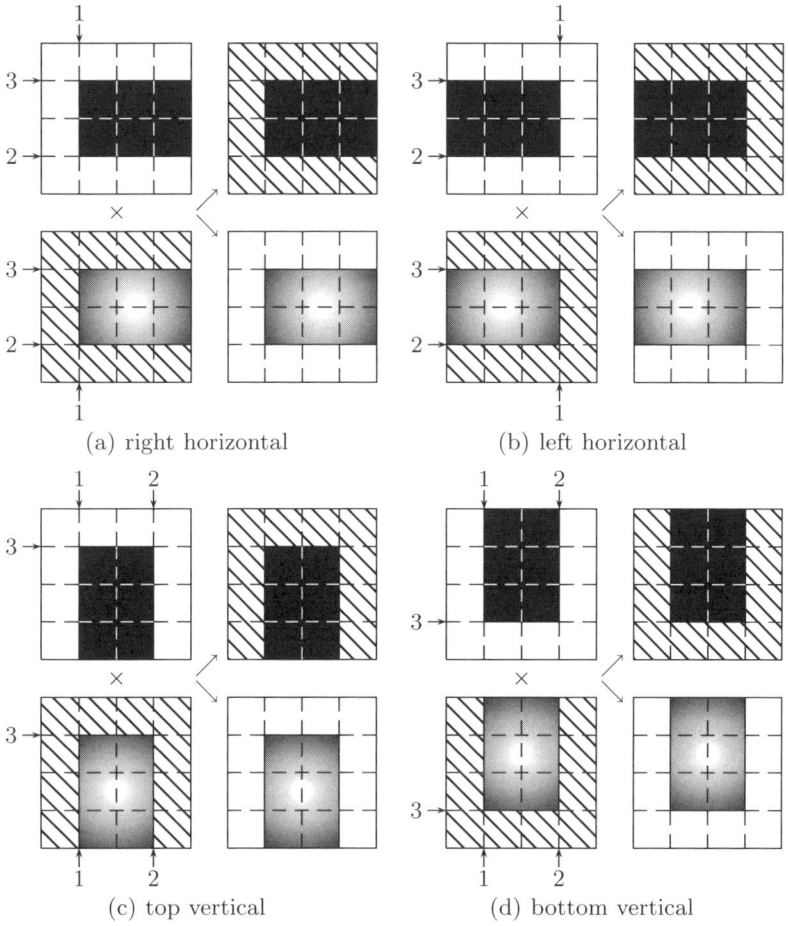

(a) right horizontal (b) left horizontal

(c) top vertical (d) bottom vertical

Fig. 3.4. Triple-point crossover of S-boxes

$$\mathcal{A}_{\mathcal{S}}^* = \max_{\beta \in \mathcal{B}^m \setminus \{0^m\}} \mathcal{A}_{\mathcal{S}_\beta(x)}, \text{ wherein}$$
$$\mathcal{S}_\beta(x) = \bigoplus_{i=0}^{i=m} \beta_i \mathcal{S}_i(x) \tag{3.9}$$

Definition 12. An S-box $\mathcal{S}$ is said to be *regular* if and only if for each $\omega \in \mathcal{B}^m$ there exists exactly the same number of $x \in \mathcal{B}^n$ such that $\mathcal{S}(x) = \omega$. The *regularity* of an S-box can be measured by $\mathcal{W}_{\mathcal{S}}^*$ defined in (3.10):

$$\mathcal{W}_{\mathcal{S}}^* = \max_{\beta \in \mathcal{B}^m \setminus \{0^m\}} \left| \mathcal{W}_{\mathcal{S}_\beta(x)} \right|, \text{ wherein}$$
$$\mathcal{S}_\beta(x) = \bigoplus_{i=0}^{i=m} \beta_i \mathcal{S}_i(x) \tag{3.10}$$

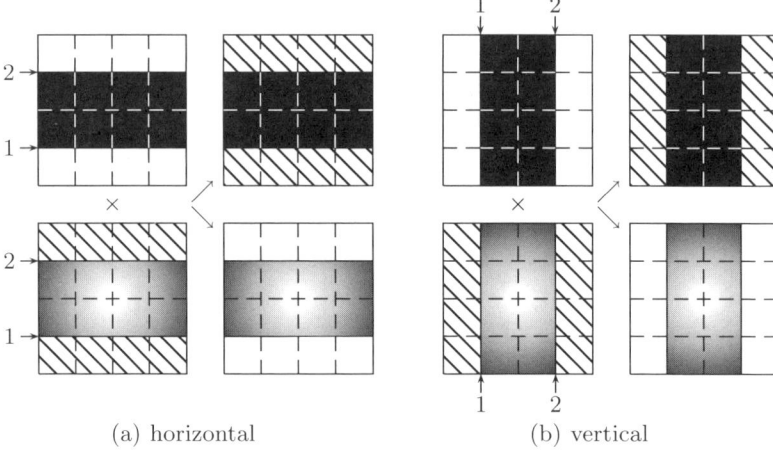

(a) horizontal   (b) vertical

Fig. 3.5. Double-point crossover of S-boxes

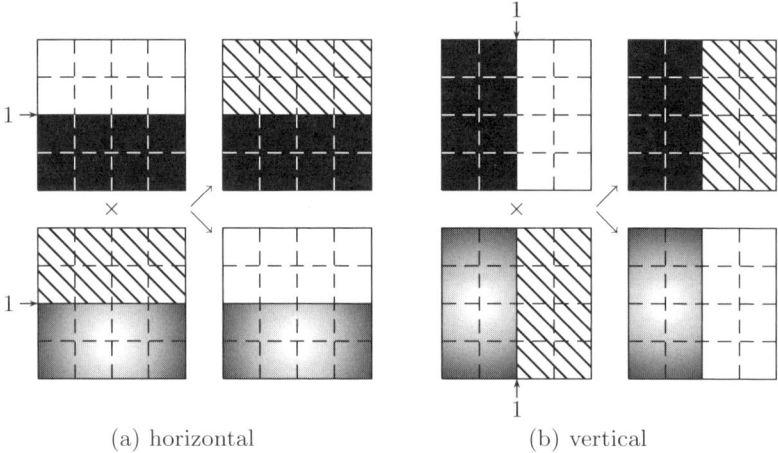

(a) horizontal   (b) vertical

Fig. 3.6. Single-point crossover of S-boxes

Note that a regular S-box $\mathcal{S}$ has $\mathcal{W}_{\mathcal{S}}^* = 2^{n-1}$.

The optimisation objectives consist of maximising regularity of the S-box as well as its non-linearity while minimising its auto-correlation. These are stated in (3.11):

$$\begin{cases} \max_{\mathcal{S}} \; \mathcal{W}_{\mathcal{S}}^* \\ \max_{\mathcal{S}} \; \mathcal{N}_{\mathcal{S}}^* \\ \min_{\mathcal{S}} \; \mathcal{A}_{\mathcal{S}}^* \end{cases} \qquad (3.11)$$

3.5 Performance Results

The Nash algorithm [3, 4], described in Section 3.3 was implemented using multi-threading available in Java^{TM}. As we have a three-objective optimisation (maximising regularity, maximising non-linearity and minimising auto-correlation), our implementation has three agents, one per objective, all three running in parallel in a computer with a Hyper-Threaded Pentium-IV processor of 3.2GHz. We used S-boxes of different size (i.e. number of input and output bits) to evaluate the performance of our approach. All the yield S-boxes were regular. The non-linearity and the auto-correlation criteria for the best S-boxes yield are given in Table 3.1 together with those obtained in [11] and [12]. For all the benchmarks, our approach performed better producing S-boxes that are *more* non-linear and *less* auto-correlated than those presented in [11] and [12]. The best solutions were always found after at most 500 generations.

Table 3.1. Characteristics of the best S-boxes $\mathcal{S}^+$ by Millan et al. [11], Clarck et al. [12] and our approach

input×output	Millan et al.		Clarck et al.		Our approach	
	$\mathcal{N}^*_{\mathcal{S}+}$	$\mathcal{A}^*_{\mathcal{S}+}$	$\mathcal{N}^*_{\mathcal{S}+}$	$\mathcal{A}^*_{\mathcal{S}+}$	$\mathcal{N}^*_{\mathcal{S}}$	$\mathcal{A}^*_{\mathcal{S}}$
8×2	108	56	114	32	116	34
8×3	106	64	112	40	114	42
8×4	104	72	110	48	110	42
8×5	102	72	108	56	110	56
8×6	100	80	106	64	106	62
8×7	98	80	104	72	102	70

In order to compare the solutions produced by the three compared approaches, we introduce the concept of *dominance* relation between two S-boxes in a multi-objective optimisation.

Definition 13. An S-box $\mathcal{S}_1$ *dominates* another S-box $\mathcal{S}_2$, denoted by $\mathcal{S}_1 \succ \mathcal{S}_2$ or interchangeably solution $\mathcal{S}_2$ *is dominated by* solution $\mathcal{S}_1$ if and only if $\mathcal{S}_1$ is no worse than $\mathcal{S}_2$ with respect to all objectives and $\mathcal{S}_1$ is strictly better than $\mathcal{S}_2$ in at least one objective. Otherwise, S-box $\mathcal{S}_1$ does not dominate solution $\mathcal{S}_2$ or interchangeably $\mathcal{S}_2$ is not dominated by $\mathcal{S}_1$.

So, at the light of Definition 13, we can see that all the S-boxes evolved by our approach dominate those yield by Burnett et al. [11]. Furthermore, the 8×4, 8×5 and 8×6 S-boxes we produced dominate those generated by Clarck et al. too. Nevertheless, the 8×2, 8×3 and 8×7 S-boxes are non-dominated.

3.6 Conclusion

In this chapter, we used a multi-objective evolutionary algorithm based on the concept of Nash equilibrium to evolve resilient innovative evolutionary S-boxes. The produced S-boxes are regular in the sense of Definition 12. Moreover, considering the dominance relationship between solution in multi-objective optimisation as defined in Definition 13, the generated S-boxes are better than those obtained by both Burnett et al. and Clark et al. This is encouraging to pursue further evolution of more complex S-boxes. Also, as a future work, we intend to apply genetic programming to evolve the resilient S-boxes as minimal Boolean equations so that the hardware of the S-boxes is given for free.

References

1. Daemen, J. and Rijmen, V., The Design of Rijndael: AES – The Advanced Encryption Standard, Springer-Verlag, 2002.
2. National Institute of Standard and Technology, Data Encryption Standard, Federal Information Processing Standards 46, November 1977.
3. Nash, J.F., Equilibrium points in n-person games, Proceedings of the National Academy of Sciences, vol. 36, pp.48–49, 1950.
4. Nash, J.F., Non-cooperative game, Annals of Mathematics, vol. 54, no. 2, pp. 286–295, 1951.
5. Kang, J.S., Yi, O., Hong, D. and Cho, H., Pseudo-randomness of MISTY-type transformations and the block cipher KASUMI, Information Security and Privacy, The 6th Australasian Conference, Lecture Notes in Computer Science, vol. 2119, pp. 60–73, Springer-Verlag, 2001.
6. Menezes A. J., Van Oorschot, P. C. and Vanstone, S. A., Handbook of Applied Cryptography, CRC Press, 1996.
7. Matsui, M., Linear cryptanalysis method for DES cipher, T. Helleseth (Ed.), Advances in Cryptology, vol. 765, Lecture Notes in Computer Science, pp. 386–397, Springer-Verlag, 1994.
8. Biham, E. and Shamir, A., Differential Cryptanalysis of DES-like Cryptosystems, Advances of Cryptology, vol. 537, pp. 2–21, Springer-Verlag, 1990.
9. Beauchamp, K. G., Walsh Functions and Their Applications. New York: Academic, 1975.
10. Dorigo, M. and Maniezzo, M., Parallel Genetic Algorithms: Introduction and Overview of Current Research, in Parallel Genetic Algorithms, J.Stender (Ed.), IOS Press, 1993.
11. Millan, W., Burnett, L, Cater, G., Clark, J.A. and Dawson, E., Evolutionary Heuristics for Finding Cryptographically Strong S-Boxes, Lecture Notes in Computer Science, vol. 1726, pp. 263–274, Springer-Verlag, 1999.
12. Clark, J.A., Jacob, J.L. and Stepney, S., The Design of S-Boxes by Simulated Annealing, New Generation Computing, vol. 23, no. 3, pp. 219–231, Ohmsha and Springer, 2005.

Industrial Applications Using Wavelet Packets for Gross Error Detection

Paolo Mercorelli[1] and Alexander Frick[2]

[1] University of Applied Sciences Wolfsburg
   Dep. of Vehicles, Production and Control Engineering
   Robert-Koch-Platz 12., 38440 Wolfsburg, Germany
   Phone: +49-(0)5361-831615 Fax. +49-(0)5361-831602.
   p.mercorelli@fh-wolfsburg.de
[2] ABB Utilities GmbH UTD/PAT2
   Kallstadter Strasse 1
   DE 68309 Mannheim, Germany
   Phone: +49 621 381 4539 Fax: +49 621 381 2244.
   alexander.frick@de.abb.com

This chapter addresses Gross Error Detection using uni-variate signal-based approaches and an algorithm for the peak noise level determination in measured signals. Gross Error Detection and Replacement (GEDR) may be carried out as a pre-processing step for various model-based or statistical methods. More specifically, this work presents developed algorithms and results using two uni-variate, signal-based approaches regarding performance, parameterization, commissioning, and on-line applicability. One approach is based on the Median Absolute Deviation (MAD) whereas the other algorithm is based on wavelets. In addition, an algorithm, which was developed for the parameterization of the MAD algorithm, is also utilized to determine an initial variance (or peak noise level) estimate of measured variables for other model-based or statistical methods. The MAD algorithm uses a wavelet approach to set the variance of the noise in order to initialize the algorithm. The findings and accomplishments of this investigation are:

1. Both GEDR algorithms, MAD based and wavelet based, show good robustness and sensitivity with respect to one type of Gross Errors (GEs), namely outliers.
2. The MAD based GEDR algorithm, however, performs better with respect to both robustness and sensitivity.
3. The algorithm developed to detect the peak noise level is accurate for a wide range of S/N ratios in the presence of outliers.

P. Mercorelli and A. Frick: *Industrial Applications Using Wavelet Packets for Gross Error Detection*, Studies in Computational Intelligence (SCI) **57**, 89–127 (2007)
www.springerlink.com   © Springer-Verlag Berlin Heidelberg 2007

4. The two developed algorithms based on wavelets (Algorithms for GEDR and peak noise level estimation) do not require the specification of any additional parameters which makes parameterization fully automated as a result. The MAD algorithm only requires the result of the peak noise level detection algorithm that has been developed and therefore is also fully parameterized.
5. The sensitivity and robustness of the algorithms is demonstrated using computer-generated as well as experimental data.
6. The algorithms work both, for off and on-line cases where in the on-line case computation times are small so that application on standard Distributed Control Systems (DCS) for large plants should be feasible.
7. A translation from Matlab to C (C++) was done using the Matlab Compiler and a COM wrapper was implemented as a functional prototype, which can be directly accessed through an excel spreadsheet.

To conclude, the developed algorithms are totally general and they are present in some industrial software platforms to detect sensor outliers. Furthermore, it is currently integrated in the inferential modelling platform of the unit responsible for Advanced Control and Simulation Solutions within ABB's (Asea Brown Boveri) industry division. Experimental results using sensor measurements of temperature, pressure and Enthalpy in a Distillation Column are presented in the chapter.

4.1 Introduction

Gross error detection and replacement (GEDR) is essential for many model-based or statistical methods in the process control field. An example would be parameter estimation (PE) and data reconciliation (DR). Here, gross error (GE) free data is assumed prior to PE and DR. Currently, only simple GEDR based on limit and rate checks is carried out within some industrial platforms for the detection of outliers. This contribution will use PE and DR as a reference example, however, one should keep in mind that other model-based or statistical methods may also be used instead. As applications in the process industries rely more and more on large amounts of raw data, which either directly originate from a distributed control system (DCS) or historians, there is an increasing need for automated data pre-processing, e.g., cleaning of data. Soft sensing, data reconciliation, and parameter estimation are example applications [13] that often require the existence of "clean data". In the case of data reconciliation, the data cleaning has been done simultaneously, see [13]. A new method of de-noising which is based on methods already introduced in [16] is presented in [17]. It is based on the comparison of an information theory criterion which is the "description length" of the data. In [14] and [15] this theoretical approach is applied. The description length is calculated for different subspaces of the basis and the method suggests choosing the noise variance and the subspace for which the description length of the data is minimum. In the work, based on [14], it is shown that the de-noising process can be done simultaneously. The approach presented in [14] and [15] is based on "minimum description length" and that's why it seems to be unsuitable for on-line detection because it needs a relatively large amount of data. Furthermore, our approach is looking for a two step de-noising technique. Simultaneous approaches often suffer from the fact that they are based on iterations which in turn impose heavy CPU load requirements on the applications. As a consequence, these approaches often become unfeasible for real-time applications and therefore have not been widely used in industry. Real-time approaches dealing with outlier detection and/or de-noising and featuring an easy means of implementation have also been developed. Historically, lower and upper limits as well as simple rate change check were used to detect outliers. More recently, in [1], [18] and [19]) the authors have developed a robust version of the Hampel filter where a lower threshold may be set in order to avoid the undesired detection (removal) of noise. The algorithm assumes knowledge of the noise level of the signal, which is often unknown. The aim of this chapter is to develop an approach that addresses outlier detection without de-noising that may be applied in real-time. This is particularly important because some applications require outlier detection without de-noising of the signal. Special attention was given to the fact that the algorithms developed may be tuned automatically during commissioning and may be easily retuned during operation. The strategy consists of using an existing Hampel algorithm (enhanced by [1]) and the newly developed algorithm, based on Wavelets, to determine robustly the noise level of the signal to be processed. The noise level was found to be the parameter of the Hampel algorithm by [1] that cannot be optimally chosen independent of the signal to be processed. The algorithm to estimate the noise is based on Wavelets. The other parameters were optimally fixed using metrics defined by means of simulation studies. Fig. 4.1 shows a block diagram of how a GEDR package (PCK) would interact with gEST, that is, how the GEDR would function on a conceptual level.

4.1.1 Modules

The following list describes the modules (blocks) and the connections in order to define the problem statement in general, the key assumptions to be made, and the implications thereof on the GEDR approaches. Dashed blocks and connections denote additional, desired features to be included on a conceptual level, however, they will not be tested or implemented at this point in time. The scope of this work is the GEDR and Quality Module (QM) where the QM has been given more consideration on a conceptual level.

- **Process and Instrumentation**: The true process values are affected/ changed by sensor (instrumentation) imperfections such as sensor aging or

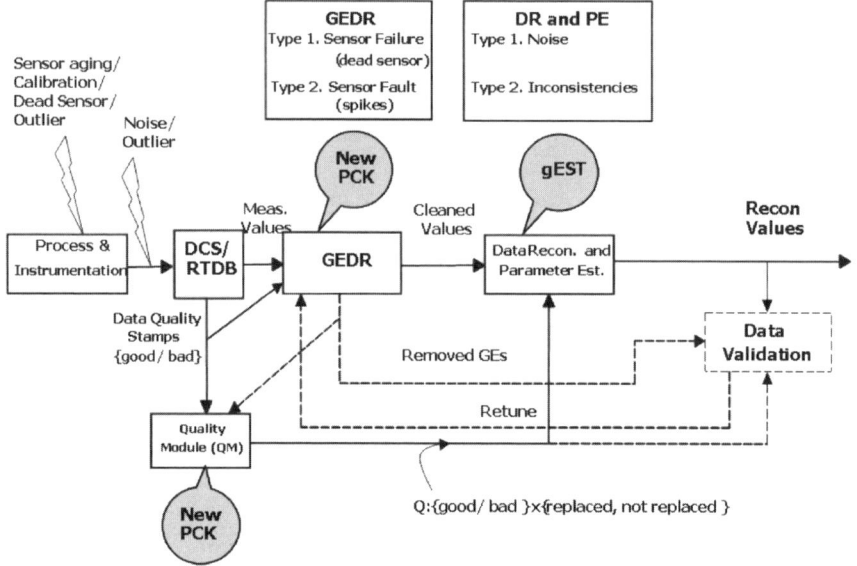

Fig. 4.1. Overview of Modules relevant to GEDR and QM

wrong/poor calibration. Often, these effects are modelled using biases. Usually, biases are assumed to be rather small as opposed to Gross Errors (GEs), which may be caused by dead sensors for example.

- **DCS/RTDB**: The DCS/RTDB logs data and stores them in a data base. The data logged consist of the values from the process and instrumentation block and additional effects: noise, outliers, and possibly interference due to the measurement; possible effects to the anti-aliasing; AD-conversion and quantization noise; and possibly LP filtering. Additional information from sensors such as quality stamps may also be available and used within the GED and the Quality Module (QM). For example, a quality stamp may say good or bad.

- **Quality Module (QM)**: Some of the functionally of the QM resides within the DCS. For example, in "2 out of 3" set-ups, where the same physical property is measured by three different sensors, the measured value with a "bad" stamp is automatically discarded and values from the other two sensors are used. Different scenarios, some of which also include soft sensors, are conceivable. The different scenarios strongly depend on the DCS system used and the configuration chosen by the engineers and operators, respectively. For our purposes, the QM module could provide additional functionality using information from the GEDR. Hence, on input the QM merely combines two streams of information. One stream comes from the DCS/RTDB and the other from the GEDR. Characteristics assigned to data points are simply merged and relayed with additional information to other blocks: DR, PE, and Data Validation. The set of characteristics generated by the DCS/RTDB may be good/bad whereas the set generated by the GEDR may be replaced/not replaced.

- **GEDR**: Gross error detection and replacement must process the data coming from the DCS/RTDB block. The GEDR cleans the data before data reconciliation (DR) and parameter estimation (PE). The GEs, which were detected and removed, should be analyzed after DR and PE to validate that they actually are GEs and in turn to retune the filter settings. GEDR replaces the GE detected with a value that fits the local behaviour of the variable. For example, the average of the last couple of the valid data points may be used to do that. This module represents the main part of our investigation.

- **Data Recon. And Parameter Est. (DR and PE)**: PE is to estimate the parameters whereas DR reduces the error between predicted and measured variables while ensuring consistent process values using redundant information. For example, if all of the mass fractions were measured they would have to add up to one, which represents redundant information. DR and PE must be done after GEDR. This can be done for both dynamic and steady state data. The DR and PE are done simultaneously within some industrial platforms. The DR is performed by introducing BIAS variables that are considered parameters. They are adjusted in order to make the error (difference between measurement and prediction) disappear. Information provided by the QM regarding the quality and possible alternative measurements should be processed within DR and PE. For example, the quality stamp may be: good/badxreplaced/non replaced whereas possible alternative measurements may be a different sensor or soft sensor. Data points identified as being replaced could be weighted differently in order to weaken their influence upon the PE and DR. Consequently, some provisions regarding the information handling/processing ability of the PE and DR should be made. These may involve, as mentioned, the possibility to change the weights of measured data points in the PE and DR. This is not possible within gEST because the weights for a variable are constant for a given experimental run, that is, no different weights may be assigned for the different data points with respect to time.

4.1.2 Gross Error Types and Examples

Two types of GEs are considered:

- Type 1 GEs, which are caused by, e.g., sensor failures and result in sustained errors in the measurements. Fig. 4.2 shows an example of a type 1 GE. It is assumed that a sensor goes dead generating a noise-free signal with value 0. The sensor failure occurs at sample number 50 and remains for 100 samples. Provisions for detection of type 1 GEs will be made. However, this will require some user interaction such as specifying the number

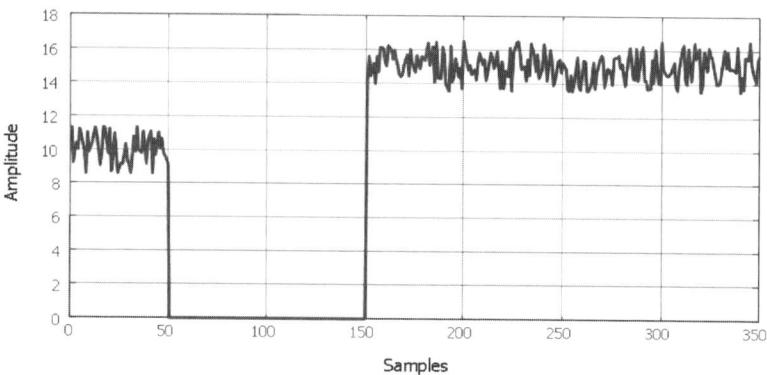

Fig. 4.2. Example of type 1 GE. Computer-generated data with dead sensor defaulting to zero at sample number 50

of constant measurements resulting in a type 1 GE. This option would make sense for variables such as analyzers for which clogging may cause such a behaviour. Henceforth, the specification such as the number of constant measurements as well as a dead band should be done by a process expert.

• Type 2 GEs, which are caused by, e.g. sensor faults, result in short, spike-like errors in the measurements

Fig. 4.3 shows data from a temperature measurement contaminated with outliers and hypothetical limits of the min and max values (horizontal thick lines) of the temperature. The limits can be used to perform an easy check to remove some of the outliers. Obviously, not all of the outliers would be removed using the min/max GE check. Another possible easy check could be implemented with a rate limitation. However, this would involve detailed process knowledge of the time constants and noise levels of the measured variables. The GEDR is to remove and replace all of the spikes shown. After detection and replacement, the cleaned data is shown in the lower part of the figure using a MAD based filter, see [1], combined together with the wavelet based peak noise level estimation algorithm. The chapter is organized as follows. Section 4.2 is devoted to the problem specification. Section 4.3 gives some background elements and presents the state of the art of the noise detection problem. At the end of the section 4.3 a new wavelet based procedure to detect noise variance is presented and validated. Section 4.4 shows an algorithm wavelet based for GED and the MAD algorithm with the noise

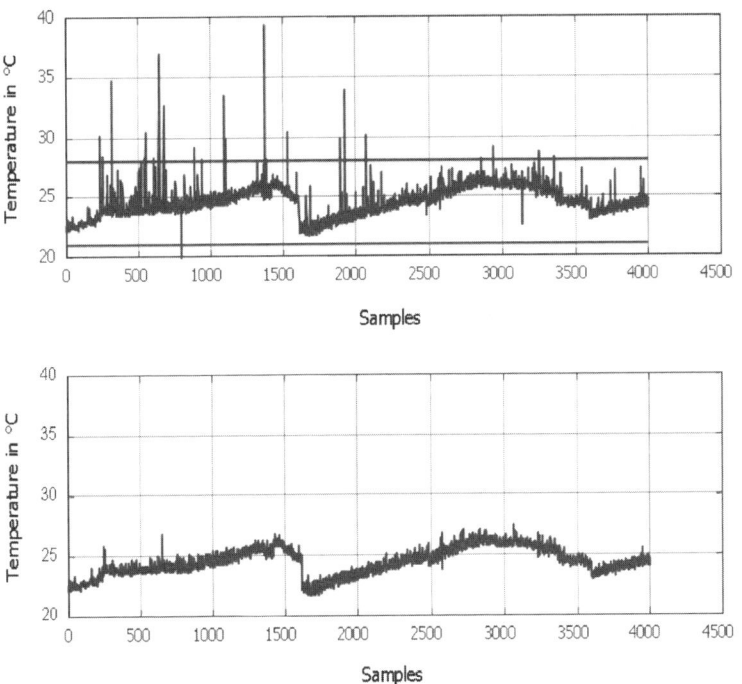

Fig. 4.3. Example of temperature measurements contaminated with outliers (top) and min, max limits. Temperature data cleaned using GEDR (bottom)

parameter estimated by the new wavelet based procedure. In section 4.5 and 4.6 results with computer simulation and real measurements are shown. The conclusions close the chapter.

4.2 Problem Specification

4.2.1 Mathematical Preliminary

Measurement errors in the context of sampled, measured variables are simply the difference between what is measured and what is the true value of the variable, that is,

$$(y)_k = (x)_k + (o)_k + (b)_k + (w)_k \tag{4.1}$$

where: $(y)_k$ is measured, $(x)_k$ the true value, $(e)_k = (o + b + w)_k$ is the error due to the measurement and k denotes the $k - th$ value of the time series. In particular o_k is the outlier, w_k is the noise and b_k represents any other kind of fault such as, for instance, a sustained error in the measurements, (see Fig. 4.2), or systematic errors due to the instrumentation. Equation (4.1) represents an additive error model as an extension of the outlier model presented in $[1, 3]$. Another error model used in the literature is a multiplicative error model where the errors are multiplied rather than added up. Often, additive errors are used to model errors occurring at inputs or outputs whereas multiplicative errors are used to model parameter errors, see [1]. Therefore, an additive error model has been chosen. The goal of the algorithms developed is to detect sensor fault related outliers. This results in a short GE that is characterized by a non-continuous, dynamic response such as an impulse or an impulse-like response. Outliers may be distinguished from noise by the amplitude of the peaks. They are short in duration, often no more than two points at a time.

Outlier Detection Problem (ODP) and Algorithm (ODA)

Therefore, the outlier index vector has been defined in the following way:

Definition 1 *For a set of data $(y)_k = (x)_k + (o)_k + (b)_k + (w)_k$ for $k = 1, ..., n$ where $(y)_k$ is measured. The number of data points n are given. The outlier detection problem (ODP) is to find $L = (l_1, ... l_n)$ for which $(o)_k \neq 0$, that is:*

$$L = \begin{cases} (l)_k = 1 & (o)_k \neq 0 \\ (l)_k = 0 & elsewhere \end{cases}$$

The index set L itself does not specify what will happen in case an outlier is actually detected. It is supposed to be a mere specification of the detection problem.

The goal of this chapter is not to prove some of the inherent features of a data cleaning filter to solve the ODP. This was done in [1]. However, in order to assess the performance of the filter implemented together with the noise detection and to simplify the further discussion some more definitions are necessary. The implementation of an outlier detection algorithm (ODA) will tag the data points within the time series $(y)_k$ that are believed to be outliers. This may be represented by the following definition of a mapping F. If the data point under consideration satisfies the criteria, as used in the Hampel filter in [1], then the map F returns a value equal to one.

Definition 2 *For a set of data $(y)_k = (x)_k + (o)_k + (b)_k + (w)_k$ for $k = 1, ..., n$ where $(y)_k$ is measured. The number of data points, n, are given. For an implementation of a filter represented by the mapping F, the index vector L^T may be computed as follows:*

$$L^T = \begin{cases} (l^T)_k = 1 & (F(y)_k) = 1 \\ (l)_k = 0 & elsewhere \end{cases}$$

The following part will summarize some of the inherent features.

4.2.2 Noise Level Detection Problem (NLDP) and Algorithm (NLDA)

Definition 3 *For a set of data $(y)_k = (x)_k + (o)_k + (b)_k + (w)_k$ for $k = 1, ..., n$ where $(y)_k$ is measured. The number of data points, n, are given. The noise level detection problem (NLDP) is to find the standard deviation for a normally distributed noise N_l. Where $N_l = \|w\|$ for $w = (w_1, w_2, ...w_n)$ and $\| \cdot \|$ is some norm. The noise actually detected will be denoted N_l^T.*

The implementation of a noise level detection algorithm (NLDA) calculates a measure of the noise of a time series $(y)_k$. The goal here must be to have an "outlier robust algorithm", that is, the NLDA should not be sensitive to the numbers of outliers present in the time series.

4.2.3 Some Remarks Regarding Wavelet Based Algorithms

Peak Noise Level and Variance

In the context of this investigation, it is referred mostly to the peak noise level estimation. This stems from the fact that the peak noise level estimator reconstructs the peak noise levels of the signal and then calculates their variance. So, when one speaks of a peak noise level estimator, the result will be sometimes close to the standard deviation of the signal and sometimes closer to the actual peak noise level. The goal was to have a robust estimator even for a large range of S/N ratios in the presence of outliers rather than a very accurate estimator for a small range of S/N ratios and no outliers. The proposed technique is wavelet based.

Wavelet Algorithm for GEDR

The developed algorithm estimates the local variance of the Lipschitz constant of the signal over a sliding time horizon. The fault (outlier) is recognized if the local Lipschitz constant lies outside the limit computed. Graphically, a flow pipe for the Lipschitz constant is constructed and if the local Lipschitz constant lies outside the flow pipe the data point is flagged and then replaced.

4.3 Wavelet Based Noise Level Determination

The MAD algorithm uses a wavelet approach to set the variance of the noise in order to initialize the algorithm. In this section this new algorithm, together with the necessary wavelet background, is presented. The developed algorithm estimates the peak noise level. A sliding time horizon is used to estimate a local noise peak. The result for each of the sliding time horizons is stored and used to determine the peak noise level. This is to make the procedure more robust. For an individual time horizon the algorithm selects appropriate wavelet coefficients that represent the noise in the system in the best way.

4.3.1 Background and State of the Art

The wavelet transform, an extension of the Fourier transform, projects the original signal onto wavelet basis functions, thus providing a mapping from the time domain to the time-frequency plane. The wavelet functions are localized in the time and frequency domain and they are obtained from a single prototype wavelet, the mother function $\varphi(t)$, by dilation and translation. Dilation may be viewed as stretching of the mother function with respect to time whereas translation is a mere shifting with respect to time. Thus, by means of dilation and translation operations performed on the mother wavelet $\varphi(t)$ a set of wavelet functions is created with the same shape as the mother wavelets yet of different sizes and locations. In this invention, digital wavelet functions are used to analyze sampled, experimental data. The sets of digital wavelet functions are referred to as wavelet packets. In general, they are characterized by three parameters, two of which may be chosen independently. The parameters depend on the number of the samples considered, that is the data window width, as well as on the Nyquist frequency (half of the sampling frequency). The wavelet packets under consideration are derived from the Haar mother wavelet: Fig. 4.4 shows a set of Haar functions. The Haar functions are iden-

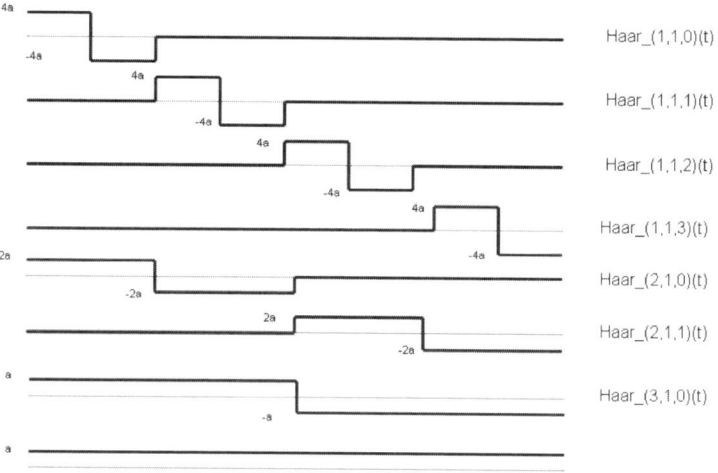

Fig. 4.4. A set of Haar functions

tified using the parameter tupel, (d, j, n). In particular, parameter j is related to the frequency shift, parameter n is related to the time shift, and additional

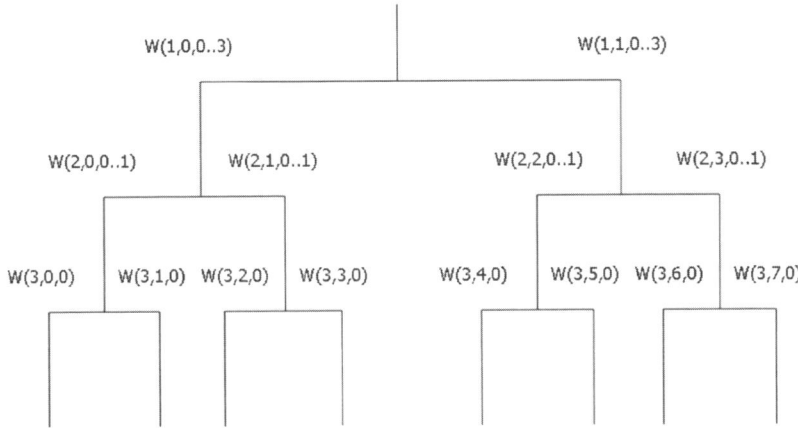

Fig. 4.5. Wavelet coefficients arranged in a tree

parameter d is useful to address the level of the tree which strictly depends on the number of the considered samples. That is, if eight samples were to be analyzed, the level of the tree would be three. In literature, wavelet functions are structured in the form of a tree such that $d = 1$ represents the highest degree of refinement with respect to time. For example, in Fig. 4.4 there are four wavelet functions at $d = 1$ whereas there are only two wavelet functions for $d = 2$. Fig. 4.5 shows the corresponding tree with the wavelet coefficients, $w(d, j, n)$, representing the contribution of each of the wavelets to the signal. The notation $w(1, 0, 0..3)$ denotes the coefficients on the first level on the very left with time shifts 0 through 3.

4.3.2 Noise Level Estimation: State of the Art

Estimation of the noise distribution density is a well-known problem from a theoretical and a practical standpoint. Contributions in this area using wavelets may be found in [2], [3], [4] and [5]. A number of different methods for selecting appropriated threshold values for wavelet de-noising have been proposed. Recent literature [6], with focus on the application domain, addresses the importance of the accurate estimation of noise. In particular, in 3 page 867 a list of threshold criteria is presented. Noise is a phenomenon that affects all frequencies and is present in different forms at different frequencies for different processes. This variability of noise generates difficulty and therefore, heuristic criteria are needed to cope with this. A heuristic used rather often is that the "true" signal tends to dominate the low-frequency band whereas the noise tends to dominate the high-frequency band. The currently used (state of the art) algorithms to estimate noise by using wavelets could be summarized as follows:

- 1 Apply the wavelet transform to a noisy signal and obtain the noise wavelet coefficients.
- 2 Threshold those elements in the wavelet coefficients which are believed to be attributed to noise.
- 3 To reconstruct the noise, apply the inverse wavelet transform to the threshold wavelet coefficients.

In this approach the crucial point is the threshold step, in other words, which wavelet coefficients are believed to be attributed to the noise. In general, wavelet thresholding methods could be classified as soft threshold as in [1], or as hard threshold [8]. Soft thresholds attribute wavelet coefficients, (which are less equal than the soft threshold given), to noise, regardless of the frequencies. Hard thresholds attribute wavelet coefficients, (which are less equal than the hard threshold given), to noise with regard to a minimum frequency given. Both thresholding methods described above may further be sub-divided into two categories: global thresholding and level-dependant thresholding. The two sub-categories relate to the level in the wavelet tree. That is, global thresholding applies only one threshold on the wavelet coefficients of one particular level whereas level-dependant thresholding applies level-dependent thresholds on all levels. The latter technique is suitable when the noise is colored or/and correlated. In [2] in 5.4 a practical example of noise estimation is given. The described algorithm uses the finest scale, that is the first level of the tree ($d = 1$, see Fig. 4.4), of wavelet coefficients to determine the noise. More specifically, the median of the wavelet coefficients at the first level of the tree is used. The use of the median as opposed to the mean makes the estimate more robust because there is a larger number of wavelet coefficients representing noise components of the signal than wavelet coefficients representing "non-noise" components of the signal. The procedure commonly used to detect the noise [2–4] is as follows:

- **Step 1** Choose the best wavelet basis representation of the measured signal by using the algorithm in [7].
- **Step 2** Decompose the signal using the selected wavelet basis.
- **Step 3** Sort the resulting coefficients and choose the threshold level as proposed in [2] or in an adaptive way on the wavelet tree.

4.3.3 The Proposed New Procedure for Peak-Noise Level Detection

The proposed procedure calculates the peak-noise level of a sampled signal. To do this, a method is proposed to represent the noise in the wavelet domain, that is, after applying a wavelet transform. The algorithm looks, at every level of the tree, for the time-frequency cell which describes at this level the incoherent part of the signal as defined in [10] and applies a property demonstrated in [10]. The property is in short the following. If the absolute value of the median over n at the time-frequency cell (d, j) (father wavelet) is not more than the absolute of the median at the time-frequency cell $(d+1, jleft)$ (left child wavelet) plus absolute of the median at the time-frequency cell $(d+1, jright)$ (right child wavelet) , then the absolute value of the father wavelet is not more than the absolute value of the median of the sum of every couple of left and right children belonging to the same branch of the tree at the level $d + n$ (all the deeper), with d and n integer.

Definition 4 *Given the oscillating part of the sequence $y(t)$ as follows*

$$y(t) - \sum_{n=0}^{2^{N_0-d}} s(d,n)\psi(d,n)(t) = \sum_{j=1}^{2^d-1} \sum_{n=0}^{2^{N_0-d}} w(d,j,n)\psi(d,j,n)(t),$$

where $s_{(d,n)} = w_{(d,0,n)}$. Then the time-frequency map of the peak values in wavelet domain is a table of real values specified $\forall d \in \mathbb{N}$ and by the triplet $(d, \hat{j}, n)$ as

$$w_{p_{(d,\hat{j},n)}} = \max_j (w_{(d,j,n)} \| \psi_{(d,j,n)}(t) \|).$$

□

Definition 5 *Given an observed sequence*

$$y(t) = x(t) + e(t).$$

Let $e(t)$ be defined as the incoherent part of the sequence $y(t)$ at every level d of the packet tree as follows:

$$e(t) = \sum_n N_{(d,\hat{j},n)} \psi_{(d,\hat{j},n)}(t),$$

where $\psi_{(d,\hat{j},n)}(t)$ are the wavelet bases and $N_{(d,\hat{j},n)}$ their coefficients. The selected wavelets are characterized by $(d, \hat{j}, n)$ indices such that:

$$\{(d, \hat{j}, n)\} = arg\left(\min_j \left(median_n \{ w_{p_{(d,j,n)}} \} \right) \right) :$$

$$\{ 0 < n \le 2^{N_0-d}, 0 < j \le 2^d - 1, \forall d \in \mathbb{N} \} \right) \quad (4.2)$$

where $median_n$ is the median calculated on the elements which are localized with index n (time localized). The coefficients $w_{p_{(d,j,n)}}$ are the wavelet table coefficients of the sequence $y(t)$ as given in Def. 4. □

It is easy to see how the idea behind Def. 5 is to sort the basis which can illuminate the difference between the coherent and incoherent part of the sequence, where *incoherent* is the part of the signal that either has no information or its information is *contradictory*. In fact, the procedure looks for the subspace characterized either from small components or from opposite components in wavelet domain.

Proposition 1 *Let $y(t)$ be a sequence and $w_{p_{(d,j,n)}}$ its corresponding wavelet sequence on different levels of the tree. At every level d and at every frequency cell j*

$$|median_n(w_{p_{(d,\hat{j},n)}})| \leq |median_n(w_{p_{(d+1,2\hat{j},n)}})| + |median_n(w_{p_{(d+1,2\hat{j}+1,n)}})|.$$

Then

$$|median_n(w_{p_{(d,\hat{j},n)}})| \leq |median_n(w_{p_{(d+m,2\hat{j},n)}})| + |median_n(w_{p_{(d+m,2\hat{j}+1,n)}})|.$$

$\forall\ m \geq 1.$ $\square$

Proof 1 *By observing that, from the orthogonality,*

$$\sum_{(d,j,n)} w_{(d,j,n)}\psi_{(d,j,n)} = \sum_{(d,j,n)} w_{(d+1,2j,n)}\psi_{(d+1,2j,n)} +$$

$$\sum_{(d,j,n)} w_{(d+1,2j+1,n)}\psi_{(d+1,2j+1,n)}. \quad (4.3)$$

Functions $\psi_{(d,j,n)}$ are organized in packets and they are scaled functions, $\forall\ d$, j and $\forall\ m \geq 1$,

$$\max\left(\psi_{(d,j,n)} \leq max(\psi_{(d+m,2j,n)}) + max(\psi_{(d+m,2j+1,n)})\right)$$

then

$$\max_j\left(w_{(d,j,n)}\psi_{(d,j,n)} \leq \max_j\left(w_{(d+1,2j,n)}\psi_{(d+1,2j,n)} + \right.\right.$$

$$\max_j\left(w_{(d+1,2j+1,n)}\psi_{(d+1,2j+1,n)} . \quad (4.4)$$

From Def. 4, $\forall\ d$ and $\forall\ n \geq 1$, it follows:

$$w_{p_{(d,\hat{j},n)}} \leq w_{p_{(d+1,2\hat{j},n)}} + w_{p_{(d+1,2\hat{j}+1,n)}}.$$

Considering that the "median" is a monotonic function:

$$median_n\left(w_{p_{(d,\hat{j},n)}}\right) \leq median_n\left(w_{p_{(d+1,2\hat{j},n)}} + median_n\left(w_{p_{(d+1,2\hat{j}+1,n)}}\right).\right.$$

Thus

$$|median_n\left(w_{p_{(w,\hat{j},n)}}\right)| \leq |median_n\left(w_{p_{(d+1,2\hat{j},n)}}\right)| + |median_n\left(w_{p_{(d+1,2\hat{j}+1,n)}}\right)|.$$

by proceeding $\forall\ m \geq 1$ the proposition is proven. $\square$

Remark 1 *In other words the proposition says that, if the minimum of the median at level d, according to Def. 4, is not more than the minimum of the median at level $d+1$ (one more in depth) then it is also not more than the minimum of the median at level $d + m$ (all the deeper), with d and m integer.*

The introduced definitions and proposition 1 allow the Building of an efficient procedure in order to estimate the level of the noise like an incoherent part of the signal. One can say that the incoherent part of a signal is the part which doesn't "match" the desired signal. The peak-to-peak noise variance is estimated taking the incoherent part of the signal defined above into consideration.

Procedure. Given a sampled signal of length n.

- **Step 0**: Specify length L as $8, 16, 32$, or 64 (dyadic length) and set the end of shifting window $esw = \mathcal{L}$. Determine the height of the wavelet tree: $h = \log_2(\mathcal{L})$. Take the last data points with respect to the end of the shifting window, "esw", of the sampled signal.
- **Step 1**: Construct the wavelet coefficient tree $W(d, j, n)$ for every $d, j > 1$ and for every n, related to the current shifting data window.
- **Step 2**: Build the wavelet coefficients corresponding to the incoherent part of the signal according to Def. 5. In particular for $d = 1$ (the first level of the wavelet tree, this yields $j = 1$).
- **Step 3**:
    - *(a)*: For all time-frequency intervals such that $j = 1, 2, ..2^d - 1$ with $d > 1$, calculate the absolute of the median for the same time-frequency interval, that is, $Wmed(d, j) = abs(median_n(W(d, j, \forall n)))$.
    - *(b)*: For every "wavelet father" $W(d, j)$ at the node (d, j) with $j > 0$ calculate its left child at the node $(d + 1, jLeft)$ and its right child $(d+1, jRight)$. Then calculate the absolute of the medians and denote them as $WmedChildLeft = abs(median_n(Wmed(d, jLeft, \forall n)))$ and $WmedChildRight = abs(median_n(Wmed(d, jRight, \forall n)))$. **While** $(d < h)$ and **For** $j =\leq 2^d - 1$. **If** $Wmed(d, j) \leq WmedChildLeft + WmedChildRight$, then denote $Peak(\hat{d}, \hat{j}) = \inf_{d,j}(Wmed(d, j))$ and limit the tree of the children of this branch **else**
- **Step 4**: Add all of $Peak(\hat{d}, \hat{j})$ of the different levels, that is, $Peak = sum_{(\hat{d}, \hat{j})}(Peak(\forall \hat{d}, \forall \hat{j}))$.

4.3.4 Validation of Peak Noise Level Estimation

The two approaches for estimating the variance have been tested.

1. Using the de-noising algorithms in the WaveLab toolbox, see [11].
2. Using the nlevelmed algorithm as an estimate for the variance.

A summary of the test is that approach 1 is not applicable and that approach 2 is a viable approach. The WPDenoise (m-file) from the WaveLab toolbox has been used, see [11], to get a cleaned signal and then reconstruct the noise by taking the difference between the cleaned and the original signal. However, this is not a sound approach. When the variance of the noise is small, the difference between the cleaned and the original signal will be dominated by the "approximation" error. When there is a high variance then much of the noise will also be present in a cleaned signal. Thus the estimated variance will only be correct for a very short rang of variances. This is illustrated in Fig. 4.6, where a sinusoidal signal with the added noise has been constructed. This was repeated for a range of variances, and the results showed that only in a small range of variances the estimator gave correct and reasonable values.

For some signals having small S/N it may be impossible to accurately estimate the variance of the noise. Fig. 4.6 shows an example, a sinusoidal function, where an inaccurate result for small S/N ratios is obtained using a procedure as the one described above.

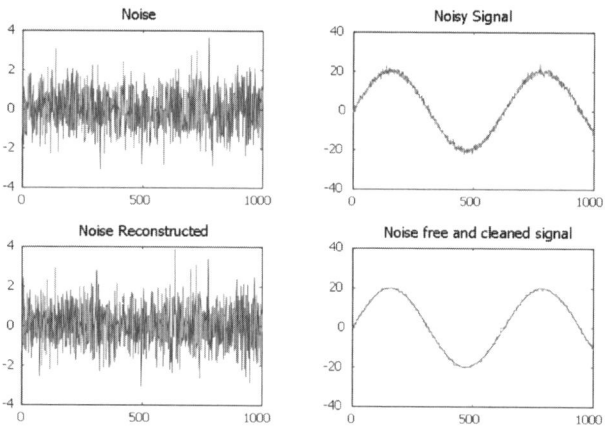

Fig. 4.6. Testing Signal

Using the WPDenoise algorithm from the WaveLab in [11] the results obtained are reported in Fig. 4.7. The same test is done with the here proposed procedure and, as it is possible to see in Fig. 4.7, the results are considerably much better.

Finally, Fig. 4.9 shows a graphical representation of the distribution of the relative error in determining the noise level error of the data sets. The noise level is determined using a wavelet-based algorithm. It almost follows a normal distribution around +10%, that is, the noise level is underestimated by 10% on average with a standard deviation of 13%. The relative error is defined as:

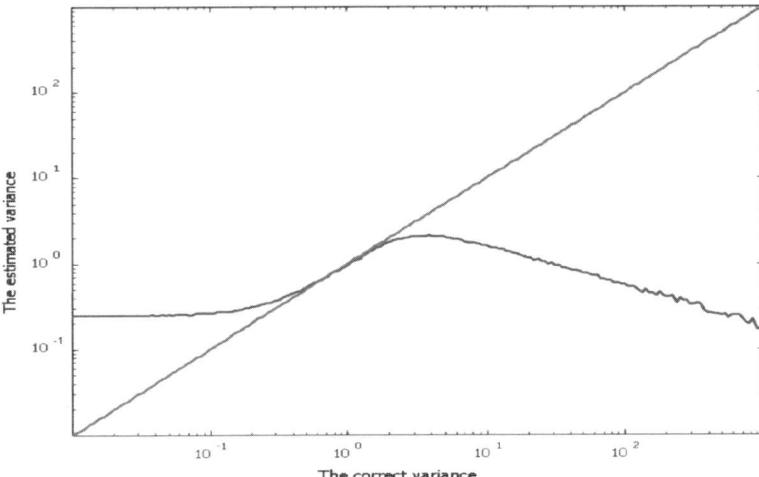

Fig. 4.7. The estimated and true variance using a threshold approach through WPDenoise algorithm from the WaveLab (the worst case)

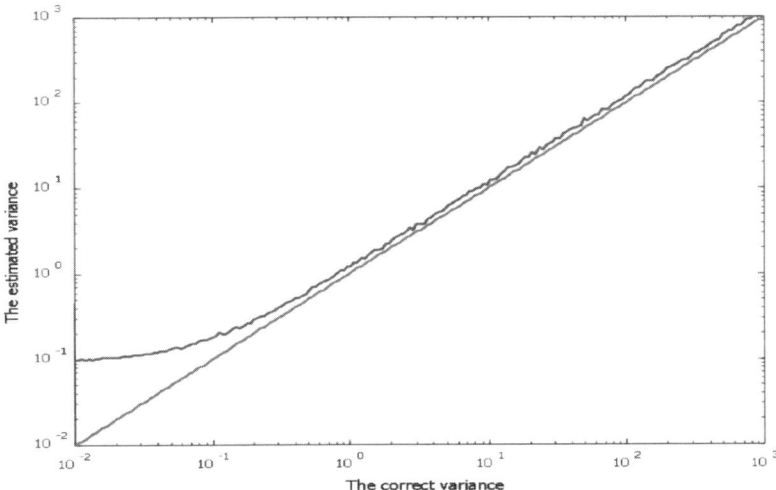

Fig. 4.8. The estimated and true variance using the here proposed procedure

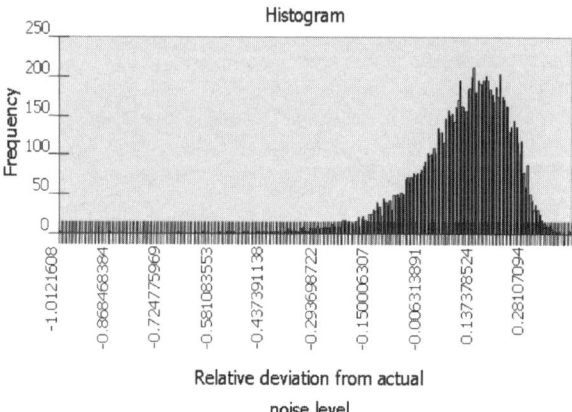

Fig. 4.9. A histogram for Noise Level Detection: the wavelet algorithm for 10000 computer-generated data sets of 200 samples each. X-axis relError as defined in the equation above. Y-axis is the density function of real Error with respect to the 10000 computer generated data sets

$$Er = \frac{nLevel - nLevelEst}{nLevel},$$

where "nLevel" is defined as follows:

- A uniformly distributed noise: "nLevel" = max amplitude of noise.
- A normally distributed noise: $nLevel = 1.5$ of standard deviation of noise.

"nLevelEst" represents the estimated noise level. The noise level detection algorithm was tested for the data sets: dryer, distillation and mining. It does a good job; the noise level always had the right order of magnitude. Fig. 4.10 shows an example of the noise level for two selected measurements in the data set. It gives an answer which is in the right order of magnitude. See also Fig. 4.7 and 4.8 where the estimated noise level is compared with the known variance of a generated noisy signal.

4.4 The Wavelet Algorithm for GEDR

The developed algorithm estimates the local variance of the Lipschitz constant of the signal over a sliding time horizon. The fault (outlier) is recognized if the local Lipschitz constant lies outside the computed boundry. Graphically,

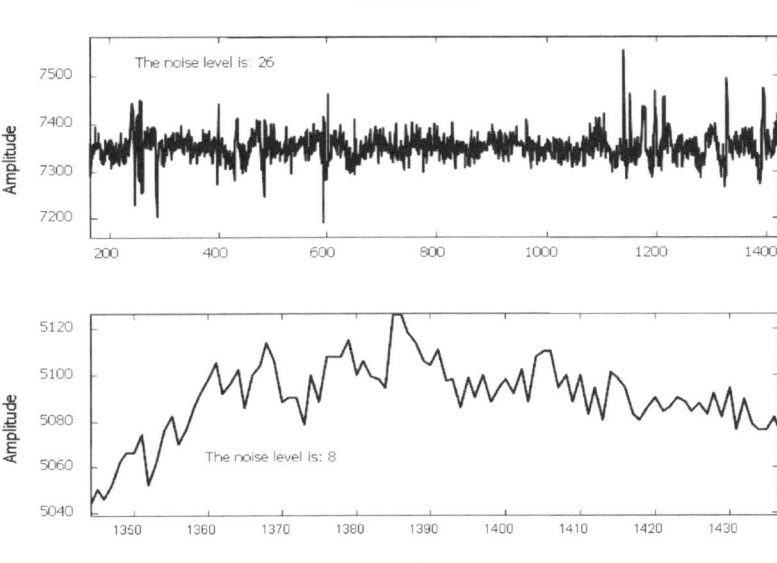

Fig. 4.10. The noise level for two selected measurements form the distillation data set

a flow pipe for the Lipschitz constant is constructed and if the local Lipschitz constant lies outside the flow pipe the data point is flagged and then replaced. One uses a sliding window with length equal to 8 samples in order to compare the results with the procedure proposed in [1]. At each sample time the window is shifted by one sample point. Decompose the signal onto the chosen Haar function. Start building the function for the first seven samples. With c_1, c_2 and c_3 one indicates the confidential constants.

- **Step 1** Calculate the standard deviation σ of the Lipschitz (L) constant for the first consecutive 7 samples and then one calculates the Lipschitz between the 7th and 8th sample.
- **Step 2** If the last Lipschitz constant is less than c_1 (Where c_1 is equal to 2), then it is not an outlier. Add this Lipschitz constant to σ. If the last Lipschitz constant is greater than c_1, then it is an outlier!
- **Step 3** Store the calculated Lipschitz constant and shift the window in order to consider the next sample of the sequence.
- **Step 4** Calculate the Lipschitz constant between the new $7^{th} - 8^{th}$ sample and add it to the stored Lipschitz.
- **Step 5** If the stored L value is less than $c_1\sigma$, then it is a single outlier and add the last Lipschitz constant to σ, see Fig. 4.11. If the stored L value is more than c_1, then check if the last two estimated Lipschitz constants have opposite signs, if no then they are multi-outliers, see Fig. 4.12. If yes,

then check if the stored L value is not less than c_2 (with c_2 equal to 3), if yes, then it is a single inverse multi-outlier, see Fig. 4.13.

If the stored L value is such that $c_1\sigma \leq L \leq c_2\sigma$, then they are multi-outliers as represented in Fig. 4.12 or Fig. 4.14.

- *(a)* In case of multi-outliers, shift the window in order to consider the next sample of the sequence and calculate the Lipschitz constant between the new $7^{th} - 8^{th}$ sample and add it to the stored Lipschitz.
- *(b)* If the stored L value is less than c_1, then it is not an outlier and add the last Lipschitz to σ. Otherwise check a heuristic safety condition on the local dynamic of the signal performed with another confidential constant c_3, if this is not verified, then they are multioutliers.

With regard to the above mentioned heuristic safety condition one can say that it is related to the dominant dynamic of the signal. In other words, one checks if the dominant dynamic of the signal before the suspected jump is close to the dominant dynamic after the jump. Fig 4.15 shows an example where this happens.

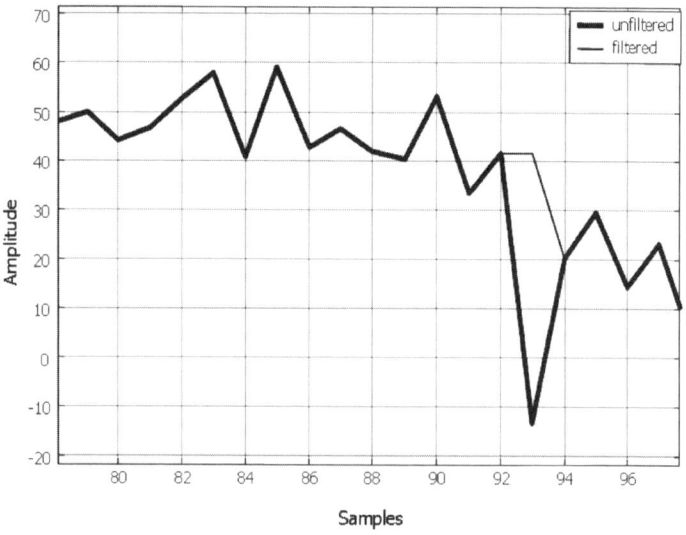

Fig. 4.11. An isolated outlier

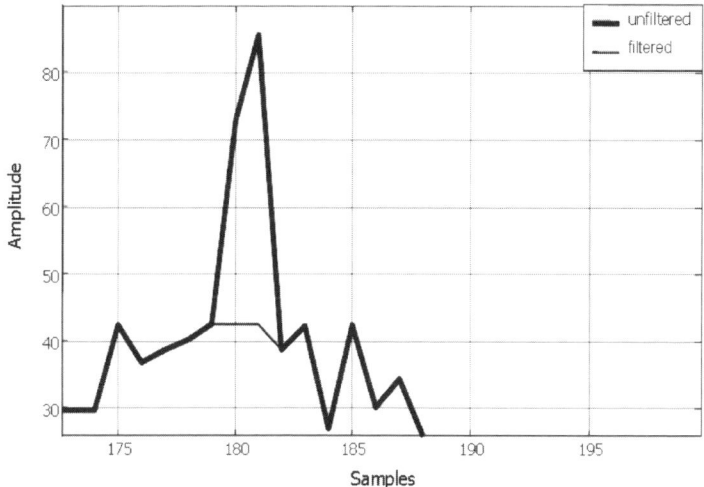

Fig. 4.12. A multiple outlier

4.4.1 Validation and Simulations

In this section one validates the procedure by using artificial data where the position of the outliers is previously known. Fig. 4.16 is related to the classical approach by using a median filter which needs to know the noise level as a priori knowledge. The results are: 94.68% of outliers correctly detected, 3.6% of outliers incorrectly detected in the data. Fig. 4.17 shows performance of the wavelet algorithm where one doesn't need prior knowledge of the noise. The results are: almost 100% of outliers correctly detected, 0% of outliers incorrectly detected in the data. It could happen in the proposed procedure that if the outliers are localized on the first part of the data the percentage of the incorrect outliers detected in the data increases and also the percentage of outliers correctly detected decreases. This is due to the initial very small standard deviation of the local Lipschitz constant during the algorithm initialization. This initialization usually takes $15 - 20$ samples. The weak points of the procedure are that it is a stochastic approach and it may not be robust in some cases. For instance, when the above mentioned standard deviation is small, the level of the percentage of the correct detection decreases. In Fig. 4.17 it is possible to see incorrectly detected outliers during the initialization of the algorithm, that is, in the first 20 samples.

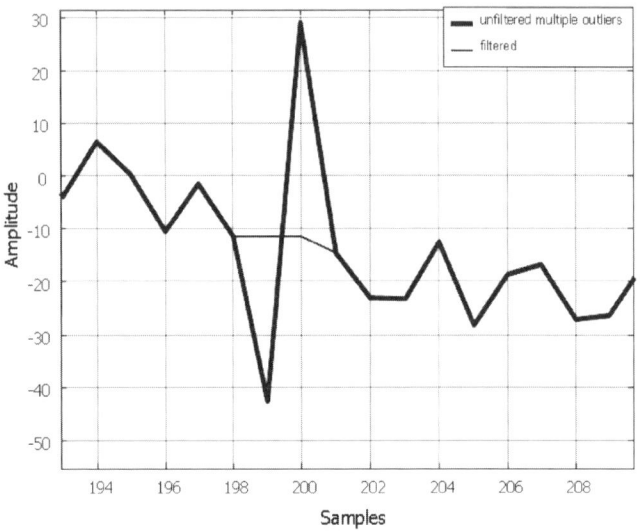

Fig. 4.13. A single inverse outlier

4.4.2 Outlier Detection Algorithm: MAD Algorithm

The algorithm itself is very simple and is described in [1] in detail. Now the original algorithm [1] and the parameters involved are briefly outlined. Basically, the algorithm uses a moving data window to determine whether or not 1) the center data point, (in the case of an off-line application), 2) or the right-most data point, (in the case of an on-line application) locally fits in the dynamics.

Fig. 4.18 shows a graphical representation of the MAD approach with the construction of the MAD flow pipe. For our tuning, the parameters were window width=8, thres=4, and var=4*nlevel. Nlevel represents the peak noise level in the data. This was determined to be the optimal setting such that, the flow pipe is wide enough to accommodate the presence of the noise, however, it is narrow enough to remove a high percentage of outliers. Also, it can be seen that the "radius" of the flow pipe is invariant under outliers and noise, which in turn makes it very robust.

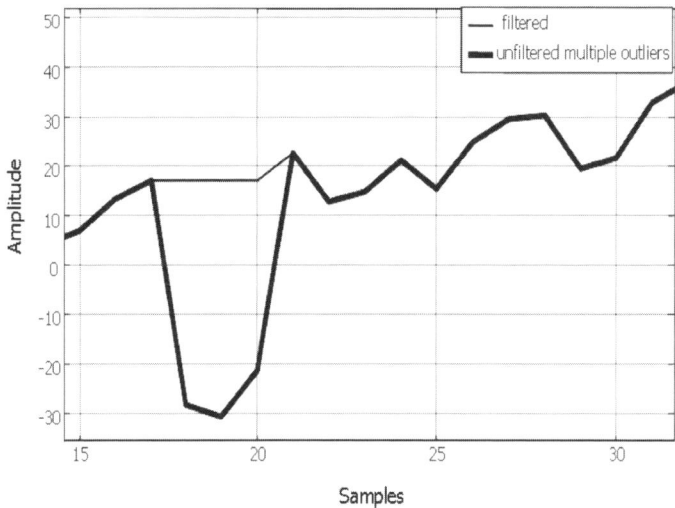

Fig. 4.14. A multiple (3) outlier

4.5 Results

Assessment Criteria The assessment criteria (metrics) are used to do two things:

1. Find optimal parameters for the MAD and the wavelet GEDR algorithms.
2. Assess the performance of the algorithms using the following metrics.
   a) Ratio of the detected and removed outliers referred to as ODR (outliers detected ratio). It can be computed both, for experimental and computer-generated data and therefore it will be the only measurement of sensitivity and robustness for the experimental data because OCDR and OIDR can only be calculated for computer-generated data.
   b) Ratio of outliers correctly detected and removed referred to as OCDR (outliers correctly detected ratio). It can be only computed for computer-generated data, where $\sharp$ of outliers which are present in data is known.
   c) Ratio of outliers incorrectly detected and removed referred to as OIDR (outliers incorrectly detected ratio). It can be only computed for computer-generated data, where $\sharp$ of outliers which are present in data is known.

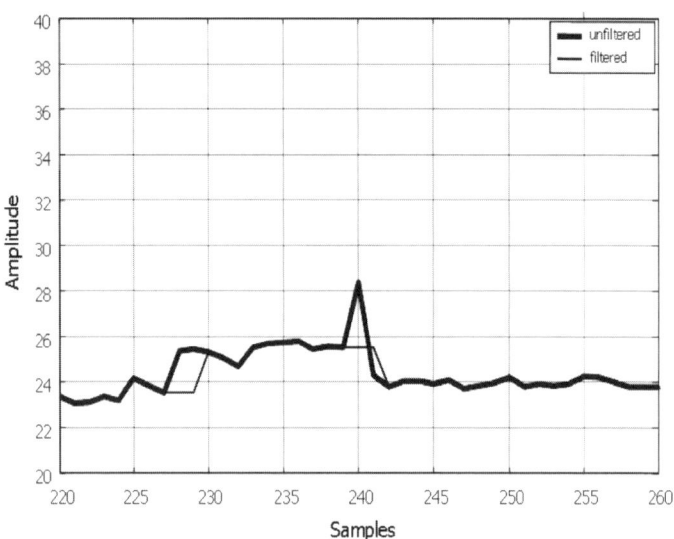

Fig. 4.15. Incorrect detection of multioutliers (on sample number 227 and 228)

d) Introduction of new data using the two-norm over all of the data referred to as SSC (sum of squared changes). It can be computed both for experimental and computer-generated data.

e) Computation Performance (CP). Both algorithms can be computed for MAD and wavelets. Note that the computation is deterministic for both approaches. In the off-line use the computation time is proportional to the number of the data points to be filtered whereas on-line it only depends on the window-width of the filter. This metric was not investigated any further because the algorithms are very fast. For example, for the on-line case using the MAD based GEDR the processing within Matlab takes approximately 1ms whereas the processing time for wavelet based GEDR algorithm is approximately 2 ms.

For the presentation of results, only OIDR and OCDR are used to optimize parametric settings for the MAD approach and the tuning of the wavelet-based approach. On the other hand, the CP is not an issue for the developed univariate signal-based approaches. The same applies to the other metrics, which were not found to be useful/critical in this context.

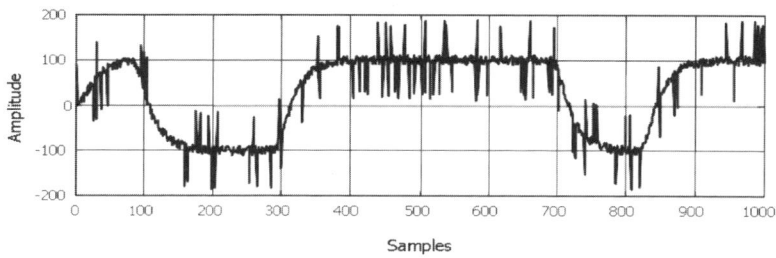

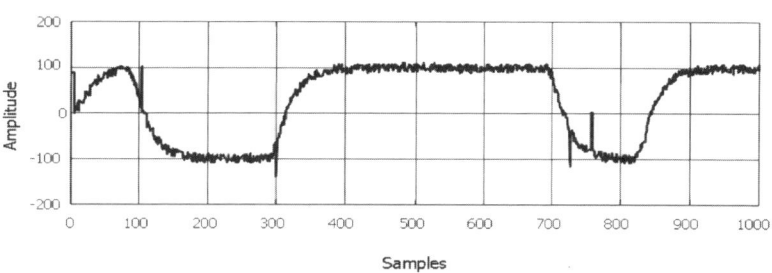

Fig. 4.16. Simulation using median filter (Algower's Algorithm) with a priori knowledge on the noise

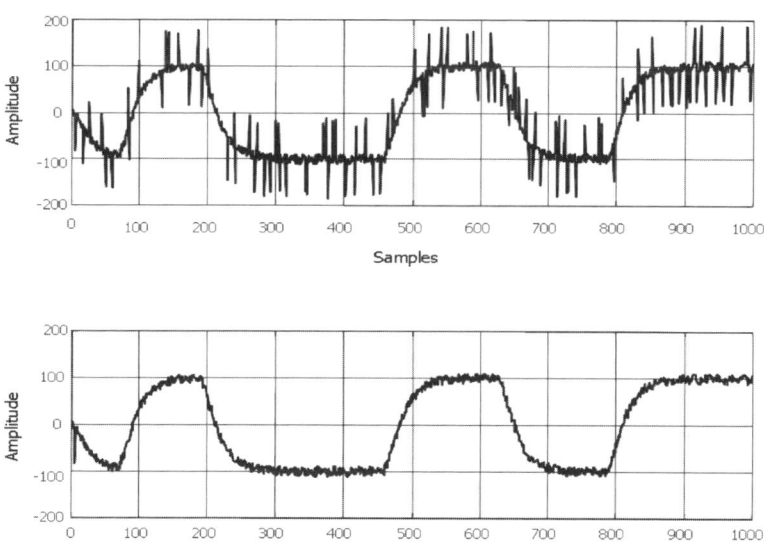

Fig. 4.17. Simulations by using wavelet algorithm without a priori knowledge on the noise

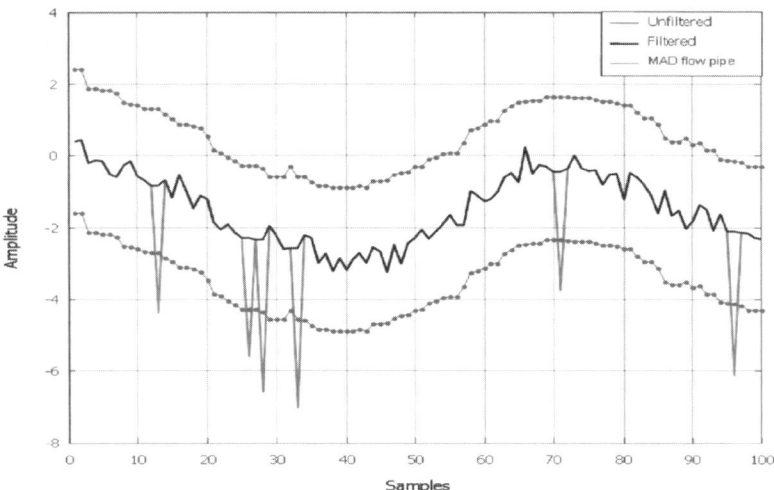

Fig. 4.18. Simulations by using wavelet algorithm without a priori knowledge on the noise

4.5.1 Algorithm Parameterization

For both wavelet approaches no parameters had to be determined. The algorithms are self-adapting and therefore do not need tuning. The goal was the generation of a reference table such that the optimal settings of the filter for the given signal parameters can be found. An optimal setting of the filter parameters was obtained by using the metrics defined above. The OIDR has been minimized while OCDR at a high level has been maintained, that is, sensitivity for robustness has been compromised. The signal parameters available for the computer-generated data are:

- Signal to Noise Ratio, SNR:= (peak to peak signal)/(peak to peak noise) as a dimensionless number. Reasonable SNRs should be between 10 and 100. For a SNR less than 10 the noise with respect to the signal becomes so dominant rendering any kind of meaningful signal processing questionable.
- Outlier to Noise Ratio, ONR:=(peak outlier)/(peak noise) as a dimensionless number. Reasonable ONRs should be greater than 1 (in this application ONRs is used greater than 4) such that they stick out compared to the measurement noise.
- Outlier Probability, OPR: The OPR defines the probability that an outlier occurs at a given time. For all practical purposes, this also represents the ratio of ($\sharp$ outliers present)/($\sharp$ data points).
- The fastest time constant ratio, DYNF:=(the fastest time constant)/(tsamp) as a dimensionless number. Reasonable values should be at least 10.

- The slowest time constant ratio, DYNS:=(the slowest time constant)/ (tsamp) as a dimensionless number. It should be at least DYNF.
- The number of transients, TRANS:= (the highest frequency of PRBS)/(the sampling frequency). The PRBS is used to generate signals, first order signals with one exponential function, second order signals with two exponential functions, and zero order responses with the PRBS as an input. Reasonable values should be less than 0.5.
- Type: uniformly and normally distributed noise.

Fig. 4.19 shows an example of computer-generated data with uniformly distributed noise. The following figures were obtained by randomly generating

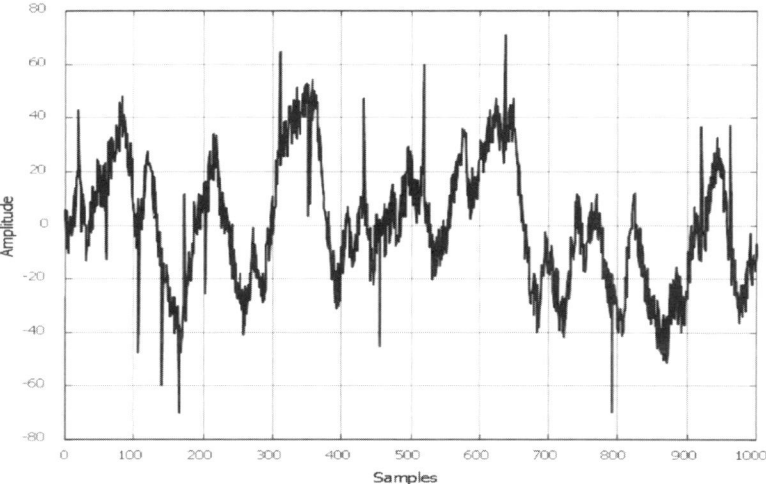

Fig. 4.19. Example of a Computer-Generated Signal: Sampling Time 1, SNR=20, ONR=4,OPR=0.02, DYNF=10, DYNS=20, TRANS=0.1

10000 data sets of 200 samples each where the above mentioned signal parameters were randomly picked out of a pre-specified range as shown in. Furthermore, the MAD filter parameters, which were optimized as presented above. Again, the wavelet algorithm does not have any parameters. Fig. 4.20 shows the accumulated distribution of OCDR using the MAD algorithm. The MAD algorithm shows a good level of sensitivity, that is, in less than 10% (y-axis) of the data sets, the algorithm of the OCDR is less than 85%. Fig. 4.21 shows similar results for the wavelet-based GEDR algorithm. However, for almost 30% of the data sets, the of ODCR is less than 85%. The worst performance is mostly due to the fast dynamics, where 5 was used as a lower limit. As a reminder 10 would be a much better ratio between a sampling rate and fastest time constant. Fig. 4.22 and Fig. 4.23 show a similar representation

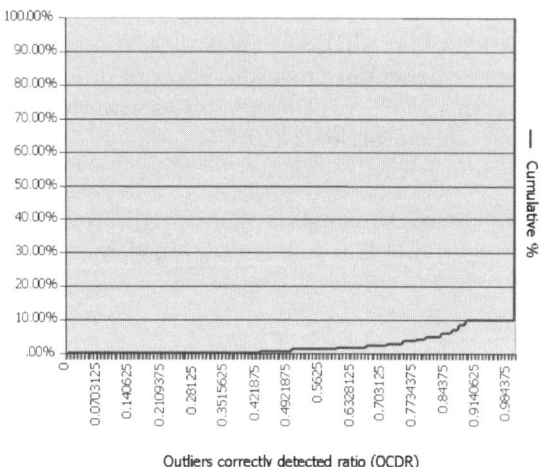

Fig. 4.20. A histogram for OCDR: the MAD algorithm for 10000 computer-generated data sets of 200 samples each. X-Axis shows the OCDR that may range from 0 (no outlier correctly detected) to 1 (all the outliers correctly detected). Y-axis show cumulative % of the data sets for which the "X-property" is true.

for the OIDR. Both algorithms show a very good robustness, that is, very few data points were misclassified. However, the wavelet based GEDR performs considerably worse. In case of the MAD based GEDR algorithm the OIDR is mostly about 99.8% of the data sets analyzed in the case of the wavelet based GEDR algorithm.

4.6 Experimental Data Sources

Data from different process industries were used to evaluate the developed algorithms. The following list represents the people that provided data with a short description of the process, domain, and the variables involved.

1. Separator Train, Oil & Gas : The separator train separates a two-phase flow of gas and liquid. Both phases contain water and oil. The measured quantities are pressures, flow rates, valve openings, and levels, and include process variables, manipulated variables, and disturbances. The variables are strongly coupled and the process shows a strong non-linear behaviour. GEs of type 2 are present in the data.

2. Dryer Section, Pulp & Paper : This process represents a dryer section within a paper mill and uses steam at different pressures for the drying.

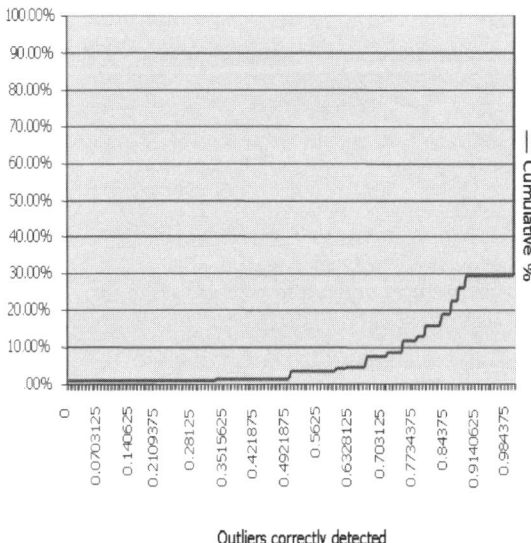

Fig. 4.21. A histogram for the OCDR: the wavelet algorithm for 10000 computer-generated data sets of 200 samples each. X-Axis shows OCDR that may range from 0 (no outlier correctly detected) to 1 (all the outliers detected correctly). Y-axis show cumulative of data sets for which the "X-property" is true.

The measured quantities are pressures, flow rates, moisture, and levels, and include process variables, set point variables, manipulated variables, and disturbances. The variables are coupled and the process shows a non-linear behaviour. GEs of type 2 are present in the data whereas type 1 GEs have not been identified yet.

3. Denox, Power Plant This process reduces the amount of NOX in an effluent stream of a boiler. The measured quantities are flow rates and compositions and include a set point variable, a manipulated variable and two disturbances. The variables are coupled, and the process shows a non-linear behaviour. No GEs are present in the data due to the "2 out of 3" measurement setup.

4. Boiler, Power Plant: This process is a boiler in a power plant. The measured quantities are pressures, temperatures, flow rates, and compositions. The variables are coupled and the process shows a non-linear behaviour. No GEs are present in the data due to the "2 out of 3" measurement set-up.

5. Mining Application: The measured quantities are flow rates.

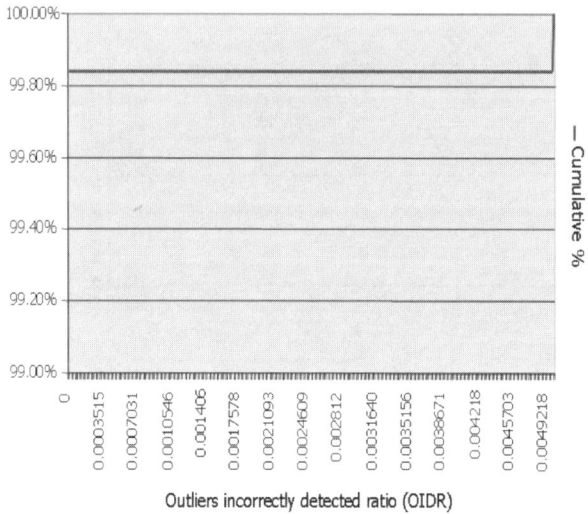

Fig. 4.22. A histogram for the OIDR: the MAD algorithm for 10000 computer-generated data sets of 200 samples each. X-Axis shows OIDR that may range from 0 (no noise misclassified as outliers) to 1 (all noise misclassified as outliers) Y-axis show cumulative % of the data sets for which the "X-property" is true

6. Distillation Column: The process represents a high-purity distillation column with 13 trays and 4 packed beds. The measured quantities are flow rates, temperatures, pressures and levels. Outliers may be present in the data. No non-linear model is available at the time.
7. Discrete Manufacturing Proprietary Information.

4.6.1 Dryer, Distillation and Mining Data with Outliers

Only the dryer data of the examined data sets were clearly contaminated outliers. There were outliers in pressure measurements, which were successfully identified both by the MAD and the wavelet filter. As it is shown in the upper plot in Fig. 4.24, the "ged_mad_filter" (MAD filter) removes the most obvious of the outliers, although there are probably some more of them. However, it does not remove any measurement noise. The "ged_wav_filter" (wavelet filter) removes more of the outliers, but it also removes some points that are clearly measurement noise. Fig. 4.25 shows an example from the distillation where it is uncertain if one is removing outliers or measurement noise. Again it is possible to see that the wavelet filter is more aggressive than the MAD one.

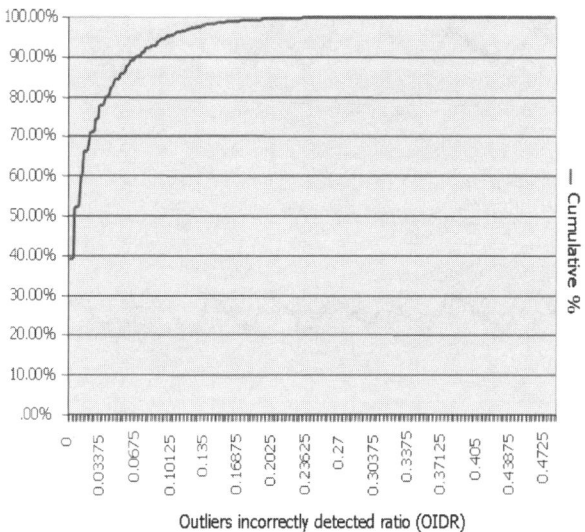

Fig. 4.23. A histogram for the OIDR: the wavelet algorithm for 10000 computer-generated data sets of 200 samples each. X-Axis shows the OIDR that may range from 0 (no noise misclassified as outliers) to 1 (all noise misclassified as outliers) Y-axis shows cumulative % of the data sets for which the "X-property" is true

Noise classified as outliers

As Fig. 4.26 shows there are cases were it is uncertain if it is a noise measurement or if it is an outlier. Fig. 4.26 shows an example where both of the algorithms remove noise in an outlier free data set.

Correlated data and outliers

Since the given algorithms are univariate such effects will not be taken into consideration. In Fig. 4.27 and Fig. 4.28 the original and the filtered data are shown. Figure shows that some outliers are detected, but only in two of the three measurements. However since this peak is present in all three of the variables and that they are strongly correlated, it is unlikely that it is an outlier. In Fig. 4.28 the same results are shown for the wavelet filter, where at least the spikes are removed in all three of the measurements.

Misclassifications

For the gross error detection algorithms to work properly it is required that the sampling period is one tenth of the "dominating" dynamics of the process.

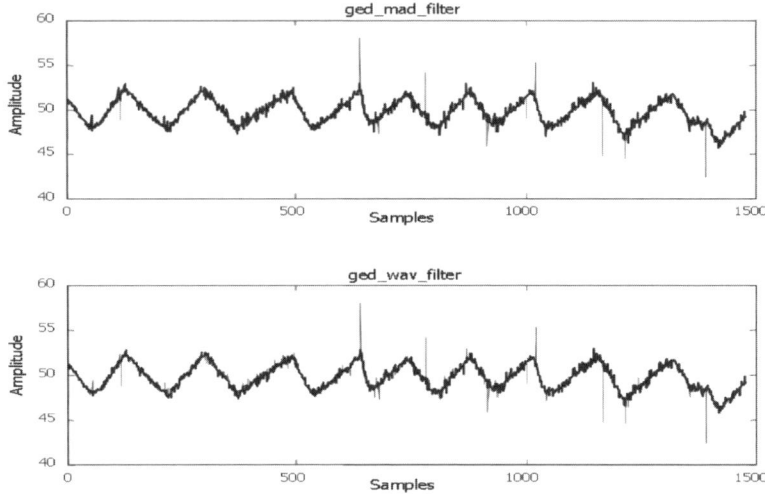

Fig. 4.24. Successful outlier detection in the dryer data set. The fine line is the original data set and the bold is the filtered one

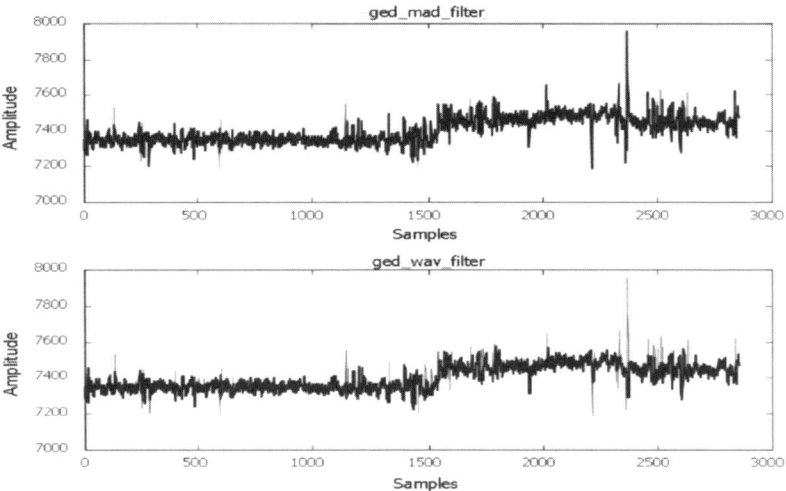

Fig. 4.25. Outlier detection. The fine line is the original data set and the bold is the filtered one

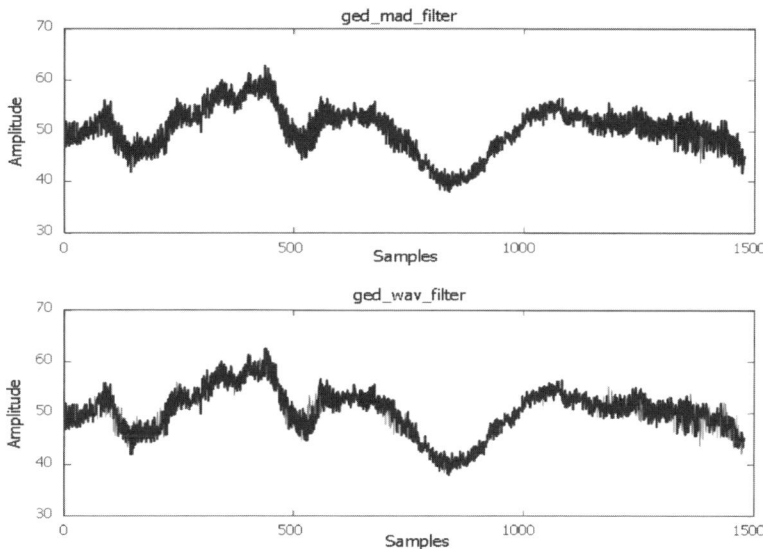

Fig. 4.26. Removal of noise. The fine line is the original data set and the bold is the filtered one

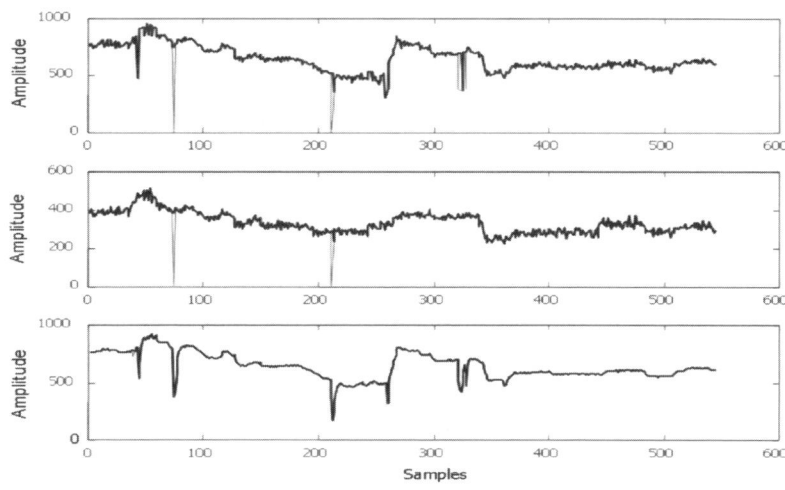

Fig. 4.27. The "ged_mad_filter" applied to the Mining data. The fine line is the original data, and the bold is the filtered one

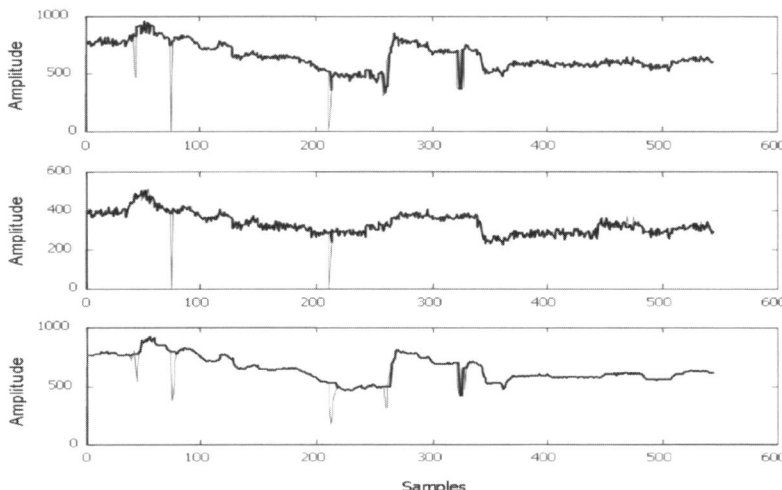

Fig. 4.28. The "ged_wav_filter" applied to the Mining data. The fine line is the original data, and the bold is the filtered one

Below, an example has been shown in which the algorithms have some problems. This may be explained in the following way: the sample time chosen is too long when compared with the dynamic of the signal. Therefore, in Fig. 4.29 the "ged_mad_filter" has problems due to the oscillating data. Fig. 4.30 and Fig. 4.31 show a problem with the wavelet approach, the signal changes are probably too fast for the filter to keep up. The stair case step is introduced due to the way the replacement is done.

4.6.2 Artificially Contaminated Data and Off-line, On-line Mode

Boiler Data and Off-line, On-line Mode

The boiler and the denox data are both outlier free. The data were used to show the effect of running the MAD based GEDR algorithm off-line and on-line. Again, the wavelet based peak level noise estimation algorithm was used to parameterize the MAD based GEDR algorithm. Fig. 4.32 shows the result obtained from the off-line mode. The top part shows raw data from the boiler where no outliers are present. The algorithm does not detect or replace any outliers, that is, the algorithm does the right thing. The same applies to the artificially contaminated data in the bottom part of Fig. 4.32 where all of the outliers are correctly detected and no misclassifications occur. Fig. 4.33 shows the result obtained from on-line mode. The top part shows raw data from the boiler where no outliers are present. The algorithm does detect and replace some outliers, that is, the algorithm misclassifies noise as outliers at around

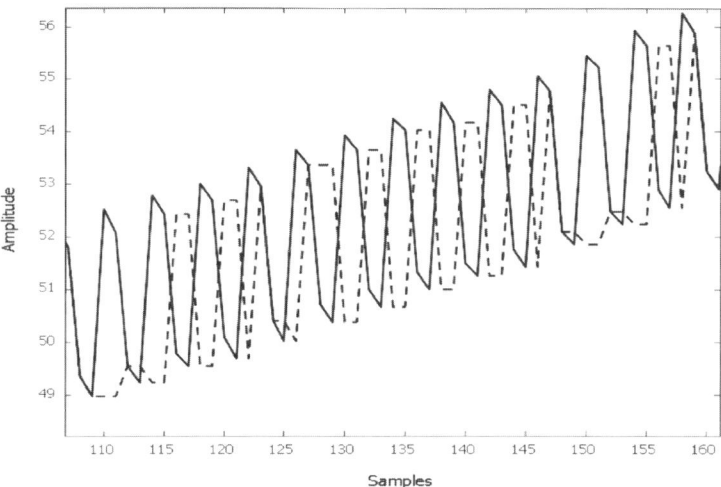

Fig. 4.29. Problems when the dynamics are near the step size for the "ged_mad_filte". The fine line is the unfiltered and the bold is the filtered one

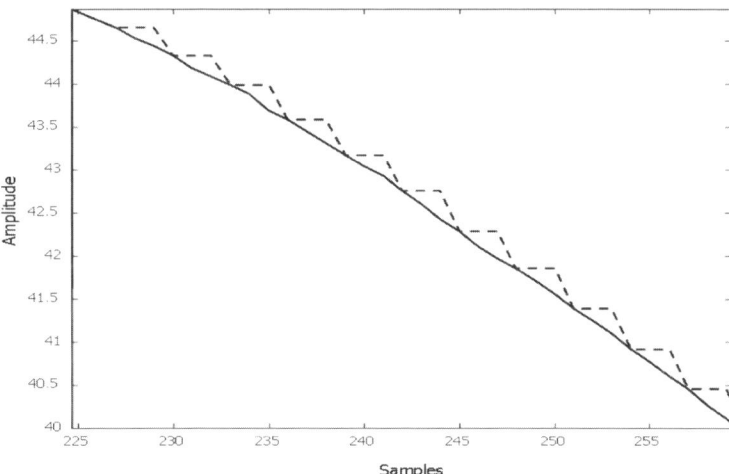

Fig. 4.30. Problems for the wavelet approach

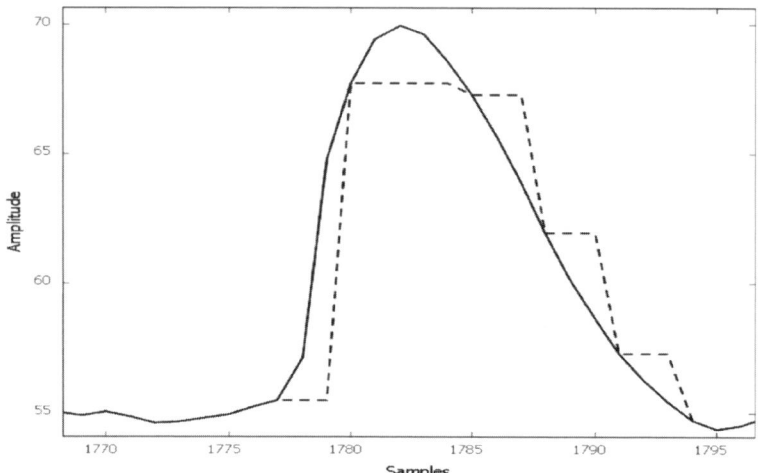

Fig. 4.31. The wavelet filter applied to the distillation case

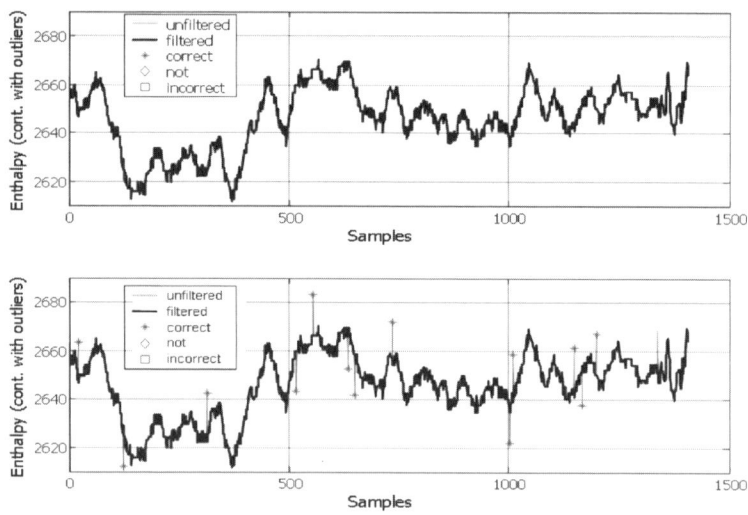

Fig. 4.32. Boiler Data Outlier Detection: The top part, raw data with no outliers due to 2-out-of-3 set-up. The bottom part, artificially contaminated with outliers and filtered using the MAD approach plus noise level detector in the off-line mode

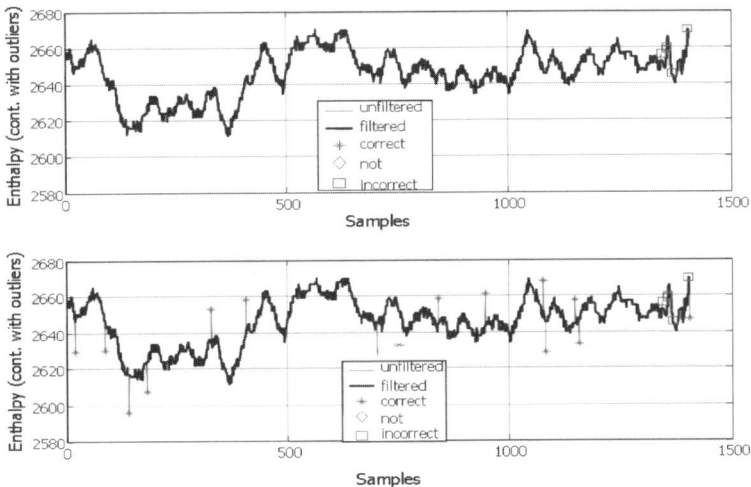

Fig. 4.33. Boiler Data Outlier Detection: The top part, raw data with no outliers due to 2-out-of-3 set-up. The bottom part, artificially contaminated with outliers and filtered using the MAD approach plus noise level detector in on-line mode

sample number 1300. The same applies to the artificially contaminated data in the bottom part of Fig. 4.33 where all of the outliers are correctly detected, however, misclassifications also occur.

Distillation Column

In Fig. 4.34, a data set from the distillation case has been contaminated with the outliers. As it can be seen the MAD algorithm performs much better than the wavelet based one. The MAD algorithm removes all the outliers and does not remove any measurements that are not outliers. On the other hand, the wavelet based approach removes some of the measurement noise and more seriously also some of the signals.

4.7 Summary, Conclusions and Outlook

The conclusions of this investigation are: MAD based GEDR algorithm together with the wavelet based peak noise level estimator:

- show very good performance with regard to robustness and sensitivity in the detected type 2 GEs, that is, outliers.

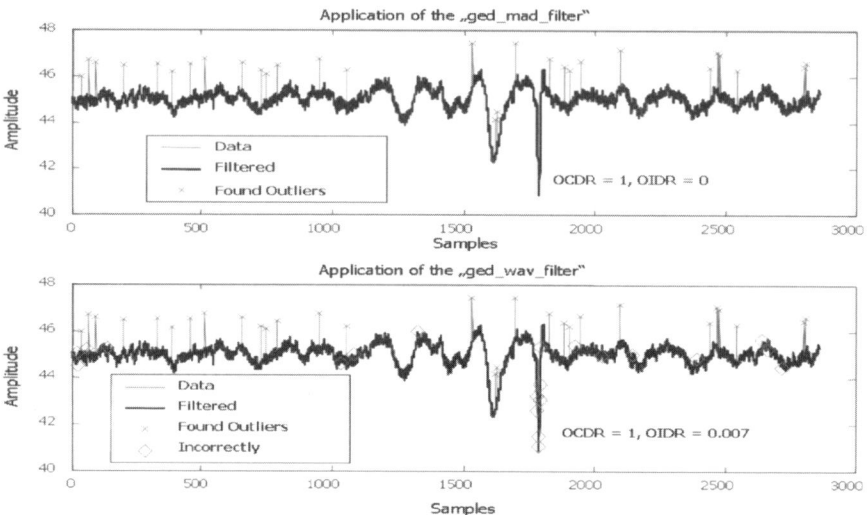

Fig. 4.34. Applications of both algorithms on an artificially contaminated data set

- show very good performance for both artificially contaminated experimental data and real contaminated experimental data.
- may be used both, for off-line and on-line applications, however, the performance disimproves in the case of an on-line application.
- The wavelet based GEDR algorithm performs well and there is still room for improvements, which should be investigated further.

References

1. P.H. Menold, R.K. Pearson, F. Allgwer (1999) Online outlier detection and removal. In: Mediterranean Control Conference, Israel
2. D. L. Donoho, I.M. Johnstone, (1994) Adapting to unknown smoothness via wavelet shrinkage. Journal of the American Statistical Association, 90 (432): 1220–1224
3. D. L. Donoho (1995) Denoising by soft thresholding. IEEE Transaction On Information Theory, 41 (3): 613–627
4. D. L. Donoho, I. M. Johnstone (1994) Ideal spatial adaptation by wavelet shrinkage. Biometrika, 81 (3): 425–455
5. D. L. Donoho, I. M. Johnstone, G. Kerkyacharian, D. Picard (1996) Density estimation by wavelet thresholding. Annals of Statistics, 24 (2): 508-539
6. R. Shao, F. Jia, E.B. Martin, A.J. Morris (1999) Wavelets and non linear principal component analysis for process monitoring. Control Engineering Practice 7: 865–879
7. R. R. Coifman, M. V. Wickerhauser (1992) Entropy based algorithm for best basis selection. IEEE Information Theory 32: 712–718

8. Rolf Isermann (1996) Modellgesttzte berwachung und Fehlerdiagnose Technischer Systeme. Automatisierungstechnische Praxis 5: 9-20

9. P. Mercorelli, M. Rode, P. Terwiesch, (2000) A Wavelet Packet Algorithm for OnLine Detection of Pantograph Vibration, In: Proc. IFAC Symposium on Transportation. Braunschweig, Germany, 13-15 June

10. Mercorelli, A. Frick (2006) "Noise Level Estimation Using Haar Wavelet Packet Trees for Sensor Robust Outlier Detection". In: Lecture Note in Computer Sciences. Springer-Verlag publishing

11. http://www-stat.stanford.edu/ wavelab/

12. I. Daubechies (1995) Ten Lectures on Wavelets. Publisher Society for Industrial and Applied Mathematics, Philadelphia (Pennsylvania)

13. J.S. Albuquerque, L.T. Biegler (1996) Data Reconciliation and Gross-Error Detection for Dynamic Systems. AIChE Journal 42: 2841–2856

14. S. Beheshti, M. A. Dahleh (2002) On denoising and signal representation. In: Proc. of the 10th Mediterranean Control Conference on Control and Automation

15. S. Beheshti, M. A. Dahleh (2003) Noise variance and signal denoising. In: Proc. of IEEE International Conference on Acustic. Speech, and Signal Processing (ICASSP)

16. R. R Coifman, M. V. Wickerhauser (1992) Entropy based algorithm for best basis selection. IEEE Trans. Inform. Theory, 32:712–718

17. R. Shao, F. Jia, E. B. Martin, A. J. Morris (1999) Wavelets and non linear principal component analysis for process monitoring. Control Engineering Practice 7:865–879

18. R.K. Pearson (2001) Exploring process data. J. Process Contr. 11:179–194

19. R.K. Pearson Outliers in Process Modelling and Identification. IEEE Transactions on Control Systems Technology 10: 55–63

5

Immune-inspired Algorithm for Anomaly Detection

Ki-Won Yeom

Information Visualization and Interaction Group
CADCAM Research Center in Korea Institute of Science and Technology
39-1 Hawolgok-dong Seongbuk-gu, Seoul, Korea
pragman@kist.re.kr

The central challenge with computer security is determining the difference between normal and potentially harmful activity. For half a century, developers have protected their systems by coding rules that identify and block specific events. However, the nature of current and future threats in conjunction with ever larger IT systems urgently requires the development of automated and adaptive defensive tools. A promising solution is emerging in the form of Artificial Immune Systems (AIS). Current AIS based Intrusion Detection Systems (IDS) have been successful on test level environments, but the algorithms rely on self-nonself discrimination, as stipulated in classical immunology. Within immunology, new theories are constantly being proposed that challenge current ways of thinking. These include new theories regarding how the immune system responds to pathogenic material. This paper takes relatively new theory: the Danger theory and Dendritic cells, and explores the relevance of those to the application domain of security. In this paper, we introduce an immune based anomaly detection approach from the abstraction of Danger theory. We also present the derivation of bio-inspired anomaly detection from the DC functionality with Danger theory, and depict two examples of how the proposed approach can be applied for computer and network security issues with preliminary results.

5.1 Introduction

The traditional prevention techniques such as user authentication, data encryption, avoiding programming errors and firewalls are used as the first line of defense for computer security. If a password is weak and is compromised, user authentication cannot prevent unauthorized use and firewalls are vulnerable to errors in configuration and ambiguous or undefined security policies as well.

K.-W. Yeom: *Immune-inspired Algorithm for Anomaly Detection*, Studies in Computational Intelligence (SCI) **57**, 129–154 (2007)
www.springerlink.com   © Springer-Verlag Berlin Heidelberg 2007

They are generally unable to protect against malicious mobile code, insider attacks and unsecured modems. Programming errors cannot be avoided as the complexity of the system is increasing and application software is changing rapidly leaving behind some exploitable weaknesses. Intrusion detection therefore becomes a necessity as an additional wall for protecting systems despite the prevention techniques.

Recently, Intrusion Detection Systems (IDS) have been used in monitoring attempts to break security, which provides important information for timely countermeasures. Intrusion detection is classified into two types: misuse intrusion detection and anomaly intrusion detection. Misuse intrusion detection uses well-defined patterns of the attack that exploit weaknesses in the system and application software to identify the intrusions. These patterns are encoded in advance and used to match against the user behavior to detect intrusion. Anomaly intrusion detection identifies deviations from the normal usage behavior patterns to identify the intrusion.

Over the last few years, Artificial Immune Systems (AIS) have become an increasingly popular computational intelligence paradigm. Inspired by the mammalian immune system, AIS seek to use observed immune components and processes as metaphors to produce systems that encapsulate a number of desirable properties of the natural immune system. These systems are then applied to solve problems in a wide variety of domains [5]. There are a number of motivations for using the immune system as inspiration for data mining; these include recognition, diversity, memory, self-regulation, dynamic protection and learning [4].

Currently, the majority of AIS encompasses two different types of immune inspired algorithms, namely negative selection (T-cell based), and clonal selection with somatic hypermutation (B-cell based) [1, 14, 19, 21].

Within immunology however new theories are constantly being proposed that challenge current ways of thinking. These include new theories regarding how the immune system responds to pathogenic material. Danger theory is one of the emerging theory that is presented by [12] and the application of this theory to AIS was identified and discussed in depth in [16]. This theory encompassed pathogenic detection, where the basis for discrimination was not centered around 'self' or 'non-self', but to the presence or absence of danger signals. It is thought that danger signals are detected and processed through professional antigen presenting cells known as Dendritic cells. Dendritic cells (DCs) are viewed as one of the major control mechanisms of the immune system, influencing and orchestrating T-cell responses, in addition to acting as a vital interface between the innate (initial detection) and adaptive (effector response) immune systems.

DCs are responsible for pathogenic recognition at the initial immune process, sampling the environment and differentiating depending on the concentration of signals or perceived misbehavior in the host tissue cells. Strong parallels can be drawn from this process to the goal of successful anomaly detection. Current anomaly detection systems frequently rely on profiling normal user behavior during a training period. Any subsequent observed behavior that does not match the normal profile (often based on a simple distance metric) classed as anomalous. At this point several alert systems are developed. However, these systems have problems with high levels of false positive errors, as behavior of users on a system changes over a period of time. Anomaly detection systems remain a high research priority as their inherent properties allow for the detection of novel instances, which could not be detected using a signature based approach. AIS featuring negative selection algorithms have been tried and tested for the purpose of anomaly detection [7].

In this paper we review the Danger theory, draw deduction of its significations, and derive an anomaly detection system using Dendritic cells combined with the theory. More specifically, we discuss the background to artificial immune systems including Danger theory immunology and Dendritic cells, and we describe a small literature review concerning the use of Danger theory and Dendritic cells (AIS) in section 5.2. Section 5.3 focuses on Dendritic cells (DCs) with respect to changing morphologies, functions, control of the immune system and outlines of bio-inspired anomaly detection mechanism in terms of the infectious non-self with danger theories. Section 5.4 shows worked examples of how the proposed approach can be used as a signal processor for early detection with complete pseudo-code, and the preliminary results is presented. Section 5.5 includes a brief analysis of the results and details of future work followed by conclusions.

5.2 Background

It is acknowledged that the Danger theory and Dendritic Cells (DCs) are a relatively new area in the realm of artificial immune systems, so to aid the reader's understanding of this paper we would like to discuss some details of both. These details have been simplified and so for a more comprehensive review of this field the reader is directed towards the literature, such as [2, 3, 6, 10, 11, 22].

5.2.1 The Danger Theory

The Danger theory attempts to explain the nature and workings of an immune response in a way different to the more traditional and widely held self/nonself viewpoint. This view states that cells of the adaptive immune system are incapable of attacking their host because any cells capable of doing so are deleted during their maturation process. This view, although seemingly elegant and generally easy to understand has come under criticism as it fails to explain a number of observations. Examples of such may be the lack of immune response to injections of inert but foreign proteins, or the failure of the immune system to reject tumors even though nonself proteins are expressed. In [12] author argues a more plausible way to describe the immune response is as a reaction to a stimulus the body considers harmful, not a simple reaction to nonself.

Danger Signals for Immune Revelation

This model allows foreign and immune cells to exist together, a situation impossible in the traditional standpoint. Matzinger hypothesizes that cellsdying unnaturally may release an alarm signal which disperses to cover a small area around that cell, Antigen Presenting Cells[1] (APCs) receiving this signal will become stimulated and in turn stimulate cells of the adaptive immune system. The term 'danger area' was coined by Aickelin and Cayzer in [11] to describe this area, in which the alarm signal may be received by APCs. This simple explanation may provide reasons for the two anomalous observations cited. Foreign proteins in the injection are not harmful and so are ignored, likewise tumor cells are not undergoing necrotic cell death and therefore not releasing alarm signals, hence no immune reaction. The nature of these alarm signals is still under discussion but some possibilities have been empirically revealed. These include elements usually found inside a cell which are encountered outside (pre-packaged signals) or chemicals such as heat shock protein, which are synthesized by cells under stress (inducible signals). As these danger signals only activate APCs, these cells in turn stimulate B and T cells into action according to the following rules:

- Signal one is the binding of an immune cell to an antigenic pattern or an antigen fragment which is presented by an APC.
- Signal two is either a 'help' signal given by a T-helper cell to activate a B-cell, or a co-stimulation signal given by an APC to activate a T-cell.

This co-stimulation signal does not fit well in the traditional self/nonself view and also leads to the question 'if a B-cell requires stimulation from a T-helper cell to become activated, what activates a T-helper cell?' As Matzinger in states "perhaps for this reason co-stimulation was ignored from its creation by Laferty and Cunningham in 1974, until its accidental rediscovery by Jenkins and Schwartz in 1986". The danger model answers this question (often referred to as the primer problem) by stating that T-cells receive their co-stimulation signals from APCs, which in turn have been activated by alarm signals.

Danger vs. Self/Non-Self

There is a criticism of the self/nonself view, which states the thymus is responsible for the negative selection of all autoreactive cells, that as the thymus

---

[1] An antigen-presenting cell (APC) is a cell that displays foreign antigen complexed with MHC on its surface. T-cells may recognize this complex using their T-cell receptor (TCR).

provides an incomplete description of self, the selection process will impart only a thymus/nonthymus distinction on T-cells. With a few simple laws concerning the described two signal activation mechanisms applied to T-cells we may provide a simple yet plausible explanation of why autoreactive cells are found in the body, yet autoimmune disease is rare:

- A resting T-cell needs two signals to be activated (as described before).
- If a T-cell receives the first signal (a binding of its receptor to an antigen) without the second signal (co-stimulation from an APC) the T-cell is assumed to have matched a host antigen and will die by apoptosis.

Thus Danger theory allows autoreactive cells to roam the body, but if that cell is to bind to a host antigen in the absence of correct antigenic presentation by an APC, the cell will die instead of becoming activated. While a number of functions related to immune response have been considered as the responsibility of the immune system, under the danger model, those are actually thought of the responsibility of the tissues. Firstly, by the second law above, simply by existing and expressing their own proteins, tissue cells induce immune tolerance towards themselves. Secondly, as an immune response is initiated by the tissues, the nature and strength of this response may also be dictated by the tissues. Therefore different types of alarm signal may result in different types of response. It has long been known that in a certain part of the body an immune response of one class may be efficient, but the same class of response in another may severely harm the host. This gives rise to a notion that tissues protect themselves and use the immune system to do so, a proposition which is in stark contrast to the traditional viewpoint whereby it is the immune system's role to protect tissues. There is still much debate in the immunological world as to whether the Danger theory is a plausible explanation for observed immune function, however we are not concerned with the plausibility of the biology. If the Danger theory is a good metaphor on which to base an artificial immune system then it can be exploited. The first steps to this are the identification of useful concepts and the application of these concepts to a suitable problem domain. It is these actions we wish to illustrate throughout the rest of this paper.

5.2.2 Dendritic Cells as Initiator of Primary Immune Response

Dendritic cells (DCs), the pacemakers of the immune response originally identified by Steinman and his colleagues representing [22], are potent antigen presenting cells (APCs) that possess the ability to stimulate naive T cells. They comprise a system of leukocytes widely distributed in all tissues, especially in those that provide an environmental interface. DCs posses a heterogeneous hemopoietic lineage, in that subsets from different tissues have been shown to posses a differential morphology, phenotype and function. The ability to stimulate naive T cell proliferation appears to be shared between these various DC subsets. It has been suggested that the so-called myeloid and lymphoid-derived subsets of DCs perform specific stimulatory or tolerogenic function, respectively. DCs are derived from bone marrow progenitors and circulate in the blood as immature precursors prior to migration into peripheral tissues. Within different tissues, DCs differentiate and become active in the taking up and processing of antigens (Ags), and their subsequent presentation on the cell surface link to major histocompatibility[2] (MHC) molecules. Upon appropriate stimulation, DCs undergo further maturation and migrate to secondary lymphoid tissues where they present Ag to T cells and induce an immune response. DCs are receiving increasing scientific and clinical interest due to their key role in anti-cancer host responses and potential use as biological adjuvants in tumor vaccines, as well as their involvement in the immunobiology of tolerance and autoimmunity [6].

Dendritic Cell Development

One of the most important findings is that DCs are not a single cell type, but a heterogeneous collection of cells that have arisen from distinct, bone marrow-derived, hematopoietic lineages. To date, at least three different pathways have been described. The emerging concepts are that each pathway develops from unique progenitors, that particular cytokine combinations drive developmental events within each pathway and that cells developing within a particular pathway exhibit distinct specialized functions [2].

The pathway of differentiation gives rise to DCs that home to peripheral tissues to take up and process exogenous Antigens (Ags) prior to migrating to the secondary lymphoid tissues to present Ags to naive T cells. Thymic DCs, on the other hand, perform a very different function being involved in the presentation of self-Ag to developing thymocytes and, hence, the subsequent deletion of autoreactive T cells. It would be appropriate for the precursors of thymic DCs to migrate to the thymus in an immature form and undergo development exposed only to self-Ag within the thymus.

---

[2] The major histocompatibility complex (MHC) is a large genomic region or gene family found in most vertebrates. It is the most gene-dense region of the mammalian genome and plays an important role in the immune system, autoimmunity, and reproductive success.

The MHC complex is divided into three subgroups called MHC class I, MHC class II, and MHC class III. The MHC class I encodes heterodimeric peptide binding proteins, as well as antigen processing molecules such as TAP and Tapasin. The MHC class II encodes heterodimeric peptide binding proteins and proteins that modulate peptide loading onto MHC class II proteins in the lysosomal compartment. The MHC class III region encodes for other immune components, such as complement components and some that encode cytokines

Immature and Maturation of DCs

DCs are derived successively from proliferating progenitor cells and non pro-liferating precursors (especially monocytes). They migrate to and reside as immature DCs at body surfaces and interstitial spaces. Immature DCs have abundant MHC II products within intracellular compartments and respond rapidly to inflammatory cytokines and microbial products to produce mature T cell stimulatory. DCs with abundant surface MHC II proteins eventually lead to apoptotic death. In most tissues, DCs are present in a so-called immature state and are unable to stimulate T cells. Although these DCs lack the requisite accessory signals for T cell activation, they are extremely well equipped to capture Ags in peripheral sites. Once they have acquired and processed the foreign Ags, they migrate to the T cell areas of lymph nodes (LNs) and the spleen, undergo maturation and stimulate an immune response.

The immature DCs have many MIICs but require a maturation stimu-lus to irreversibly differentiate into active T cell stimulatory, namely mature DCs. [13] has described an *in vitro*[3] system involving monocytes reverse trans-migrating across an endothelial monolayer that offers a possible explanation. This type of situation would occur when cells move from tissues to afferent lymph. It is possible that veiled DCs in lymph originate from monocytes in tissue that interact with the lymphoid endothelium to acquire the properties of immature DCs. If the monocytes also phagocytose particles before they reverse transmigrate, then the cells become typical mature DCs; the process occurs within 48 hours. The cells posses several markers that are expressed by mature DCs but are weak or absent in other leukocytes. Dendritic cells that have matured from monocytes in this in vitro system also express very high levels of surface MHC class II and CD86[4] and, in complete contrast to monocytes.

Maturation of DCs is crucial for the initiation of immunity. This process is characterized by reduced Ag-capture capacity and increased surface express-ion of MHC and co-stimulatory molecules. However, the maturation of DCs is completed only upon interaction with T cells. It is characterized by loss of phagocytic capacity and expression of many other accessory molecules that interact with receptors on T cells to enhance adhesion and signalling (co-stimulation). Expression of one or both of the co-stimulatory molecules on the DCs are essential for the effective activation of T lymphocytes and for

---

[3] This type of research aims at describing the effects of an experimental variable on a subset of an organism's constituent parts. It tends to focus on organs, tissues, cells, cellular components, proteins, and/or bio-molecules. Overall, it is better suited for deducing the mechanisms of action.

[4] CD86 is a co-stimulatory molecule constitutively expressed by human antigen presenting cells which interacts with CD28 and CTLA-4 expressed by T cells.

IL-2[5] production. These co-stimulatory molecules bind the CD28[6] molecules on T lymphocytes. If this fails to occur at the time of Ag recognition by the TCR (T Cell Receptor), an alternative T lymphocyte function may result, namely induction of anergy [3].

5.3 IDS based on Danger Theory and DCs Properties

In the body the immune system may only become activated within the danger area, likewise we may only need to activate antibodies within some area of interest in an AIS. One of other advantages may be to harness the role of the tissues and assign different responses to different contexts. We have given thought to the implementation of such ideas, although with no such system having been produced to date we have no literature to refer to for guidance. Based on the biology of the Danger theory, we may identify a set of Danger theory characteristics. There are also a number of desirable characteristics exhibited by DCs that we want to incorporate into an algorithm. In order to achieve this, the essential properties, i.e. those that heavily influence immune functions, have to be abstracted from the biological information presented. From this we produce an abstract model of DC interactions and functions, with which we build our algorithm.

5.3.1 Properties of DCs for IDS

DCs are examined from a cellular perspective, encompassing behavior and differentiation of the cells and ignore the interactions on a molecular level and direct interactions with other immune system cells. DCs have a number of different functional properties that we want to incorporate into an algorithm. Bearing this in mind, we can abstract a number of useful core properties, listed below and represented graphically in Figure 5.1 and Figure 5.2.

DCs' first function is to instruct the immune system to act when the body is under attack, policing the tissue for potential sources of damage. DCs can perform a number of different functions, determined by their state of maturation. Modulation between these states is facilitated by the detection of signals within the tissue - namely danger signals, PAMPs (pathogenic associated molecular patterns), apoptotic signals (safe signals) and inflammatory cytokines which are described below. The characteristics of the relevant signals are summarized below as described in [8,17]

---

[5] Interleukin-2 (IL2) is an interleukin, or hormone of the immune system that is instrumental in the body's natural response to microbial infection and in discriminating between foreign (nonself) and self.

[6] The antigen CD28 is one of the molecules that provide co-stimulatory signals, which are required for T cell activation.

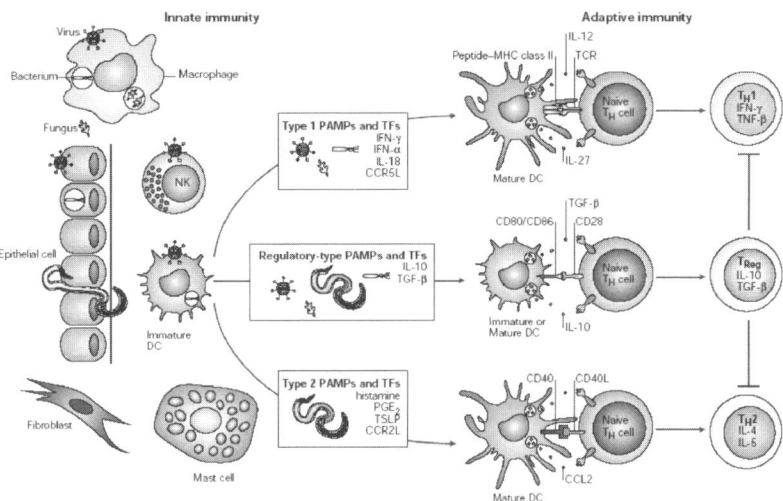

Fig. 5.1. Dendritic-cell polarization is influenced by the type of microorganism that is recognized and the site of activation [18]

- PAMPS are pre-defined bacterial signatures, causing the maturation of immature DCs to mature DCs through expression of 'mature cytokines'.
- Danger signals are released as a result of damage to tissue cells, also increasing mature DC cytokines, and have a lower potency than PAMPs.
- Safe signals are released as a result of regulated cell death and cause an increase in semi-mature DC cytokines, and reduce the output of mature DC cytokines.
- Inflammatory cytokines are derived from general tissue distress and amplify the effects of the other three signals but are not sufficient to cause any effect on immature DCs when used in isolation.

DCs reside in tissue where they are classed as immature DCs (ImmDCs). Whilst in tissue, DCs collect antigen (regardless of the source) and experience danger signals from necrosing cells and safe signals from apoptotic cells. Maturation of DCs occurs in response to the receipt of these signals. On maturation, DCs exhibit the following behavior: collection of antigen ceases; expression of co-stimulatory molecules (necessary for binding to powerful T-lymphocytes) and chemical messengers known as cytokines; migration from the tissue to a lymphatic organ such as a lymph node; and presenting antigen to T-lymphocytes.

As shown in figure 5.2, the context of the tissue (i.e. the type of signals experienced) is reflected in the output chemicals of the DC. If there is a greater concentration of danger signals in the tissue at the time of antigen collection, the DCs will become fully mature DCs with license (LmatDCs), and will express LmatDCs cytokines. Conversely, if the DC is exposed to safe signals, the

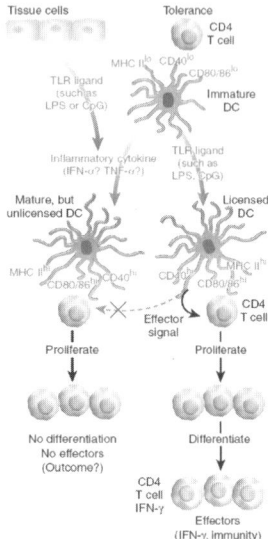

Fig. 5.2. Indirect inflammatory signals cause DC maturation but not license DCs to drive CD4 T-cell effector functions [8]

cell matures differently becoming a unlicensed mature DC, expressing ULmat-DCs cytokines. The LmatDCs cytokines activate T-lymphocytes expressing complimentary receptors to the presented antigen. Any peripheral cells expressing that antigen type are removed through the activated T-lymphocyte. The ULmatDCs cytokines suppress the activity of any matching T-cell, inducing tolerance to the presented antigen. The context of the antigen is assessed based on the resulting cytokine expression of the DC [18].

5.3.2 Abstraction of Anomaly Detection Algorithm

In our model, a combination of infectious non-self model with Danger theory is used to investigate an artificial anomaly detection algorithm. As shown, the orchestration of an adaptive immune response via DCs has many subtleties. Only the essential features of this process are mapped in the first instance as we are interested in building an anomaly detector, not an accurate simulation.

Input data is generated through monitoring environments, behavior of user or system call, and context information by several predefined scan sensors. The collection of input data is also transformed into antigen and signals. Input signals are combined with some information such as data length, name or ID, or process/program service ID etc. This is achieved through using a population of artificial DCs to perform aggregate sampling and data processing. Using multiple DCs means that multiple data items in the form of antigen are sampled multiple times. If a single DC presents incorrect information, it becomes inconsequential provided that the majority of DCs derive the correct context.

The sampling of data is combined with context information received during the antigen collection process. Different combinations of input signals result in two different antigen contexts. Unlicensed mature antigen context implies antigen data was collected under normal conditions, whereas a mature antigen context signifies a potentially anomalous data item.

Our data and method of processing is very different from other AIS, which rely on pattern matching of antigen to drive their systems. In our algorithm, antigen is only used for the labeling and tracking of data, hence we do not have a similarity metric. The representation of the antigen can be a string of either integers or characters. Signals are represented as real-valued numbers, proportional to values derived from the context information of the data-set in use. For example, a danger signal may be an increase in CPU usage of a computer. The value for the CPU load can be normalized within a range and converted into its real-valued signal concentration value. The signal values are combined using a weighted function (Equation 1) with suggested values of the weights derived from empirical data based on immunologists' lab results

The nature of the response is determined by measuring the number of DCs that are fully mature. The closer this value is to 1, the greater the probability that the antigen is anomalous. The value is used to assess the degree of anomaly of a given antigen. By applying thresholds at various levels, analysis can be performed to assess the anomaly detection capabilities of the algorithm.

$$C_{[csm,LmatDCs,ULmatDCs]} = \frac{((W_P * C_P) + (W_S * C_S) + (W_D * C_D))}{W_P + W_S + W_D)} * \frac{(1 + IC)}{2}$$

$$(5.1)$$

As stated in [15, 17], we are treating DCs as processors of both exogenous and endogenous signal processors. Input signals are categorized either as PAMPs (P), Safe Signals (S), Danger Signals (D) or Inflammatory Cytokines (IC) and represent a concentration of signal. They are transformed to output concentrations of co-stimulatory molecules (csm), ULmatDCs cytokines and LmatDCs cytokines. The signal processing function described in Equation 5.1 is used with the empirically derived weightings presented in Table 5.1.

Additionally, Safe Signals may reduce the action of PAMPS by the same order of magnitude. Inflammatory cytokines are not sufficient to initiate maturation or presentation but can have an amplifying effect on the other signals present. This function is used to combine each of the input signals to derive values for each of the three output concentrations, where C_x is the input concentration and W_x is the weight. These weightings are based on unpublished biological information and represent the ratio of activated DCs in the presence and absence of the various stimuli e.g. approximately double the number of DCs mature on contact with PAMPs as opposed to Danger Signals.

Inflammatory cytokines are not sufficient to initiate maturation or presentation but can have an amplifying effect on the other signals present. This function is used to combine each of the input signals to derive values for each of the three output concentrations, where C_x is the input concentration and W_x is the weight.

Each DC transforms each value of using the following equations with suggested values for weightings given in equation 5.2 and 5.3. Both the equations and weights are derived from observing experiments performed on natural DCs.

Unconditional Packet Transforming (UPT)

$$LFP = \frac{packets\,actually\,forwarded}{packets\,to\,be\,forwarded}$$
$$= \frac{\#^L_{(m;M)} - \#^L_{([m];M)}}{\#^L_{(M;m)} - \#^L_{(M;[m])}} \tag{5.2}$$

over a sufficiently long time period L.

Selective Packet Transforming (SPT)

$$SPT = \frac{packets\,from\,sources\,actually\,being\,forwarded}{packets\,from\,sources\,to\,be\,forwarded}$$
$$= \frac{\#^L_{([s];m;M)}}{\#^L_{([s];M;m)} - \#^L_{([s];M;[m])}} \tag{5.3}$$

over a sufficiently long time period L.

We now describe some notations of statistics:

- $\#_{(\&;m)}$ - the number of incoming packets on the monitored node m.

- $\#_{(m;\&)}$ - the number of outgoing packets from the monitored node m.

- $\#_{([m];\&)}$ - the number of outgoing packets of which the monitored node m is the source.

- $\#_{(\&;[m])}$ - the number of incoming packets of which the monitored node m is the destination.

- $\#_{([s];M;m)}$ - the number of packets that are originated from s and transmitted from M to m.

- $\#_{([s];M;[m])}$ - the number of packets that are originated from s and transmitted from M to m, of which m is the final destination.

Table 5.1. Weights for the signal processing function

Weight factors	csm	ULmatDCs	LmatDCs
PAMPs(P)	2	0	2
Danger Signals	1	0	1
Safe Signals	2	3	-3

In nature, DCs sample multiple antigens within the same section of tissue. To mirror this, we create a population of DCs to collectively form a pool from which a number of DCs are selected for the sampling process, in a similar manner to [17]. An aggregate sampling method should reduce the amount of false positives generated, providing an element of robustness. For such a system to work, a DC can only collect a finite amount of antigen. Hence, an antigen collection threshold must be incorporated so a DC stops collecting antigen and migrates from the sampling pool to a virtual lymph node. In order to achieve this we will use a fuzzy threshold, derived in proportion to the concentration of co-stimulatory molecules expressed.

In order to add a stochastic element, this threshold is within a range of values, so the exact number of antigens sampled per DC varies in line with the biological system. On migration to the virtual lymph node, the antigens contained within an individual DC are presented with the DC's maturation status. If the concentration of mature cytokines is greater than the ULmatDC cytokines, the antigen is presented in a 'mature' context. It is possible to count how many times an antigen had been presented in either context to determine if the antigen is classified as anomalous. In order to crystallize these concepts, a worked example and details of a basic implementation are given in the next section.

5.4 DCs based Implementation of Practical Applications

To illustrate the use of this immune inspired approach to the artificial immune systems, we introduce two cases such as Denial-of-Message (DoM) attack and port scanning. The DoM nowadays is frequently attacked by malicious user using included bomb script, vicious contents, or embedded virus and port scan attack to search whether there is any vulnerable port is used as the first step of intrusion and manifold attacks affecting crucial effects on the computer system. In following subsection we describe how our research is able to adopt to those cases more detail.

5.4.1 A Detection of DoM Attack

We propose a more secure sampling system which an algorithm by which a broadcasting base station detects the failure of nodes to receive its broadcasts. We presume the attackers goal is to deny the broadcast to as many nodes as possible while remaining undetected by the base station. Our goal, then, is to limit the attacker's capacity to achieve this, so that the attacker's increased disruption of broadcasts results in an increased probability that he is detected, irrespective of the strategy he pursues. During this section we use a number of small sections of pseudocode to illustrate our example. In this pseudocode we use AB to refer to an initially empty set of artificial antibodies.

Signals and Antigen

The system we propose works over two distinct stages. The first will be an initialization and training stage with the second as a running stage. During this second stage the system will detect a disruptive adversary and in the meantime, reduce the number of acknowledgments sent to the base station, by having a subset of recipients acknowledge each broadcast, where this subset is computed deterministically but in a way that is hidden from the attacker. During the first stage, a summary of which is given in pseudocode 1, the system must generate an initial collection of antibodies, and so for a given amount of time the system may observe user actions when confronted with a new broadcast message. If the cryptographic mechanisms we employ cannot be broken by the attacker, then the attacker can ascertain whether an uncompressed node should acknowledge a broadcast only by observing the node produce the acknowledgment. This, of course, is too late for the attacker to disrupt the node from receiving this broadcast. Moreover, if the attacker disrupts the acknowledgment, then this provides evidence of his presence to the base station.

```
PROCEDURE InitializationOfTrain()
   While (size of AB ¡ a threshold)
      If (user expresses DISINTEREST in an message item from net-
work packets)
            create new antibody (ab) from broadcast message
            add newly generated antibody (ab) to AB
            For (ab ∈ AB)
               clone and mutate ab to maximize affinity with ab
               add best n clones to AB
```

Pseudocode 1. Initialization and Training

When the repertoire of immune cells has reached a given size the system may run on new data as described in pseudocode 2. The AIS will convert all incoming message into a format such that affinity between it and antibodies can be evaluated. Conceptually therefore the broadcast message is an antigen. The system makes a distinction between the terms antigen and message, as follows. Antigen is the name used to refer to a processed message which contains just a generalized representation of the original destination and the class assigned to the message. The system starts initializing an antigen count to 0. When this count reaches a certain number (K) of antigens, we use the latest K antigens to perform clonal selection and mutation with the set of antibodies as described in the Update_Population procedure.

Note that it is important that this procedure is performed only after we have a reasonable number of duly processed antigens, to avoid antibodies adapted to just one antigen and preserving generality. One of the main advantages of the immune inspired approach is this built in ability to adapt. The use of the clonal selection principle here has the effect of allowing our antibody set to change over time. The clonal selection procedure will lead to an increase in the size of set AB over time and so to counter this the final line of this procedure will remove the w most unhelpful antibodies in the set AB.

Detection algorithm

The danger signal should signal something is wrong but there is still time to recover, and should come from something the antigens have little or no control over. There is also no reason why the danger signal must indicate danger in the real world; it could be a positive signal as long as it signals that something of significance to the system is taking place in a particular area. In this instance we considered several possibilities for such a danger signal, such as abnormal frequency and/or size of messages. Although the system might work with a combination of danger signals, each of them requiring a somewhat different response, in this paper we focus on a single danger signal based on the idea that network packet has too many messages. Hence, this part of the pseudocode works as follows. First, the system computes the degree of danger. If the system has no messages waiting to be scan then we may not have a danger signal at all. If however there are many messages a danger signal may be raised. Every time a new message arrives the danger area should be reevaluated based on the current state of the network. Although in the natural immune system the danger area is spatial, in our system we are not so constrained.

The nature of this danger area must be decided upon based on what we want it to signal or how we want to react to it. Hence in this example we propose a temporal danger zone, the size of which will vary according to a measured value of the danger signal. Thus the messages in the packets which may have been let through previously may become candidates for removal on receipt of a danger signal. This temporal danger area will therefore stretch into the past and make the system even more adaptive: the larger the degree of danger, the larger the size of the danger area. Once the danger area has been computed, the system has to decide whether the message is abnormal or not.

This decision consists of predicting the class of the network, based solely on the affinity between each message and the antibodies. Hence, the system computes the affinity between each message in the danger area and each antibody.

```
PROCEDURE Continuous_adaptation()
  antigen_count ← 0
  LOOP
    receive incoming message from network packets
    ag ← preprocess message into antigen
    antigen_count ← antigen_count + 1
    IF(antigen_count = K)
      AG ← last K antigens
      Update_Population(AG)
      antigen_count ← 0
    compute degree of danger()
    WHILE(danger is high)
      compute temporal danger zone
      AG ← all message in the danger zone
    FOREACH(ag ∈ AG)
      FOREACH(ab ∈ AB)
        compute affinity (ab,ag)
      high_aff ← highest affinity value
      IF(high_aff > a threshold)
        remove the network packet

PROCEDURE Update_Population(AG)
  FOREACH(ab ∈ AB)
    FOREACH(ag ∈ AG)
      compute affinity(ab,ag)
      U ← (ag | ag's class is uninteresting)
      I ← (ag | ag's class is interesting)
      quality_ab ← ∑_{i∈U} aff(ab, ag_i) − ∑_{j∈I} aff(ab, ag_j)
    clone and mutate ab in proportion to quality_ab
    remove from AB the w antibodies with the lowest value of
quality_ab
```

Pseudocode 2. More secure mail adaptation

For each message in the danger area, if the affinity between that message and the most similar antibody is greater than a threshold, then the message is considered uninteresting (means that it is normal network packet or message). Otherwise the message is considered interesting (means it is malicious message) and remove the packet from the network. This is an analogy to the natural situation as a recognition of an antigen by an antibody is signal one. There is a correspondence in the natural system between the release of a danger signal and the activation of APCs which in return supply the co-stimulatory signal to activate T-cells. Just the presence of danger in this context may therefore take the place of signal two.

For clarification and to illustrate that by our definition this system is truly danger inspired we may draw the parallels described in Table 5.2.

Table 5.2. Parameters for secure e-mail

Description	Values
Tissue	Ethernet packet receiver
Danger Signal 1	High affinity between antibody and antigen(message)
Danger Signal 2	Receipt of danger signal
Source of danger	High number of undefined messages
Danger area	Temporal
Immune response	Move message to temporary storage
Localized response	Only messages in the danger area are considered to be remove from temporary storage

5.4.2 Experiments and Results

To achieve the desired realistic behavior, we have performed all of our simulations with radio noise accumulation. We use the default values for the remaining network device parameters. We find that the teachability behavior of our simulated broadcasts is similar to the behavior observed in experiments conducted on actual sensor nodes. For each data point, we use at least 20 simulation runs.

At the beginning of each simulation run, we allow sufficient time for the table exchanges necessary in this distance-vector protocol to stabilize. During each individual experiment, the topology remains constant, minimizing the need to send routing updates.

For each experiment, we first randomly generate a network topology. Then, we ran simulations with no attackers to derive the parameter, as defined in Equation 5.1. Note that parameters obtained in this way will account for the natural loss rate, including any message loss due to collision or contention from the acknowledgment routing protocol overhead. After obtaining parameters, we ran many simulation rounds with varying percentages of randomly compromised nodes. Each compromised node sends back an acknowledgement if sampled, but compromised nodes do not forward any broadcast messages.

The general analysis of the optimal attacker behavior demonstrates the tradeoffs an attacker makes when trying to achieve his goals. In particular, finding the optimal set of nodes the attacker should compromise is intimately tied to a particular network topology. Our analysis on the proposed approach is based on the number of deprived nodes, rather than the number of attacking nodes. This serves to decouple our detection results from the attacker model and broadcast protocol we used. For our experiments, the network density was rather low. Few nodes had more than five neighbors within radio communication range. This case benefits the attacker, since a commonly considered goal of the attacker is to create a vertex cut set of compromised nodes, partitioning the network.

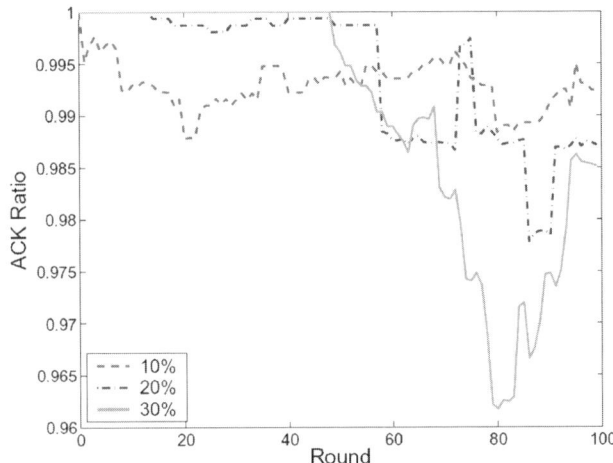

Fig. 5.3. The ratio of received to expected acknowledgements computed at the base station

Results and Analysis

In this section, we study the impact of the DoM attack on broadcast protocols. We are interested in the case where the attacker has compromised sufficiently many nodes to have a significant impact on how many legitimate nodes receive the broadcast message. Since flooding is the most general and widely used broadcast algorithm, we consider flooding in this section. As a sufficient number of nodes are compromised, their impact becomes evident. As the number of victim nodes in mid-sized networks (hundreds of nodes) increases, the attackers impact on the network becomes detectable by SIS. First, we present a naive example to convey the intuition behind detection. Figure 5.3 illustrates the ratio of received to expected acknowledgments as computed at the base station.

After the 49th broadcast, some percentage of nodes are randomly chosen to be malicious. The impact on the ratio of received to expected acknowledgements is evident, as is the unique level of natural loss for each topology. Thus, an effective threshold that can be used to detect an attack is dependent on the natural loss characteristics of the particular topology, and configuring a scheme to detect attackers in sensor network broadcast is best done post-deployment.

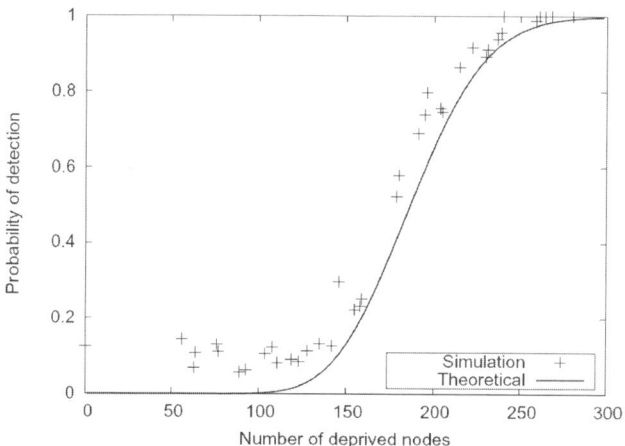

Fig. 5.4. The Probability of detection of the DoM attack based on DCs

Figure 5.4 illustrates the detection capabilities of our approach in simulation, and compares those results with the theoretically expected performance, based on the parameters (defined in Equation 5.1) observed in 0-attacker broadcast rounds. We performed the experiment on a simulated network containing 600 nodes. We used the same randomly generated topology for multiple simulation runs, while varying which nodes were malicious. Figure 5.5 depicts the false positive rate we observed. The x-axis represents, the theoretical false positive rate. is used to pick the threshold of detection. The y-axis shows the simulated false positive rate (the number of false positives over 2000 rounds of broadcast to 599 nodes).

5.4.3 A Detection of Port Scan Attack

Port scanning is a technique for discovering hosts' weaknesses by sending port probes. Although sometimes used by system administrators for network exploration, port scanning generally refers to scans carried out by malicious users seeking out network vulnerabilities. The negative effects of port scans are numerous and range from wasting resources, to congesting the network, to enabling future, more serious, attacks.

While a port scan is not an intrusion per se, it is a 'hacker tool' used frequently during the information gathering stage of an intrusion. This can reveal the topology of a network, open ports and machine operating systems.

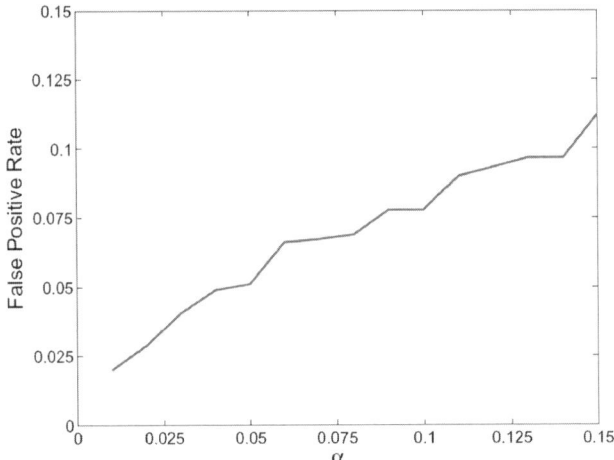

Fig. 5.5. Simulated false positive rate

The behavior of outgoing port scans provide a small scale model of an automated attack. The algorithm is applied to the detection of an outgoing port scan across a range of IP addresses, based on the ICMP 'ping' protocol.

There is a plethora of tools that aim to determine a system's weaknesses and determine the best method for an attack. The best known and documented tool is nmap. Nmap uses a variety of active probing techniques and changes the packet probe options to determine a host's operating system. Nmap offers its users the ability to randomize destination IPs and change the order of and timing between packets. This functionality can obscure the port scanning activity and thus fool intrusion detection systems. Other port scanners include queso, checkos, and SS. However, these tools do not provide all the capabilities of nmap and thus are not as popular. In this second case study we use the nmap to simulate the port scan attacks.

Signals and Antigen

Data from the monitored system are collected for the duration of a session. These values are transformed into signal values and written to a log file. Each signal value is a normalized real-number, based on a pre-defined maximum value. For this experiment the signals used are PAMPs, danger and safe signals. Inflammatory cytokines (Si4) do not feature as they are not relevant for this particular problem. PAMPs are represented as the number of 'unreachable destination' errors. When the port scan process scans multiple IP addresses indiscriminately, the number of these errors increases, and therefore is a positive sign of suspicious activity. Danger signals are represented as the number of outbound network packets per second. An increase in network traffic could imply anomalous behavior. The safe signals in this experiment are the inverse rate of change of network packets per second. This is based on the assumption that if the rate of sending network packets is highly variable, the machine is behaving suspiciously. None of these signals are enough on their own to indicate an anomaly. In these experiments the signals are used to detect the port scan (Table 5.3).

Table 5.3. Parameters for detecting port scanning

Description	Values
Tissue	Monitored data-set
Danger Signal 1	PAMP, unreachable destination
Danger Signal 2	The number of outbound network packets
Safe Signal	The inverse rate of changes of network packets
Source of danger	High network traffic
Antigen	Process ID
Immune response	Detection of port scan events
Localized response	Protect the scanned port

Detection Algorithm

In this case we apply the similar procedure as described in 5.4.1. Signals, though interesting, are inconsequential without antigen. To a DC, antigen is an element which is carried and presented to a T-cell, without regard for the structure of the antigen. Antigen is the data to be classified, and works well in the form of an identifier, be it an anomalous process ID or the ID of a data item. At this stage, minimal antigen processing is performed and the antigen presented is an identical copy of the antigen collected. Detection is performed through the correlation of antigen with signals. Following pseudocode shows the procedure of detection of port scan.

5.4.4 Experiments and Results

For the purposes of our analysis, we define a port scan as all anomalous messages sent from a single source during the trace period. To test the local warning system, we use a worm model developed in the packet level network simulator - Georgia Tech Network Simulator [20]. Because this is a real packet level simulator (which includes network router congestion, TCP latency, and other stochastic events) every run takes time. A small network was simulated with a realistic packet level model instead of using a partial analytical model.

```
PROCEDURE InitializationOfStatus()
  While (size of AB < a threshold)
    If (detector expresses NOSYMTOMS in network)
      create new antibody (ab) from log file
      add newly generated antibody (ab) to AB
      For (ab ∈ AB)
        clone and mutate ab to maximize affinity with ab
        add best n clones to AB

PROCEDURE Continuous_adaptation()
  antigen_count ← 0
  LOOP
    receive incoming network packets
    ag ← preprocess packets into antigen
    antigen_count ← antigen_count + 1
    IF(antigen_count = K)
      AG ← last K antigens
      Update_Population(AG)
      antigen_count ← 0
    compute degree of danger()
    WHILE(danger is high)
      compute temporal danger zone
      AG ← all packets in the danger zone
    FOREACH(ag ∈ AG)
      FOREACH(ab ∈ AB)
        compute affinity (ab,ag)
      high_aff ← highest affinity value
      IF(high_aff > a threshold)
        block the scanned port
```

Pseudocode 3. Initialization and Detection

Our motivation for using packet level network simulation is to validate our local early warning system and the results of our analytical models in an Internet-like setting. We used a hybrid network topology with clustering backbone and hierarchical sub-networks. The address space was populated with uniformly random hosts. Then, the vulnerable hosts were selected from the population in a uniform random way. Since the simulated network is small compared to the whole Internet, we set the time required to detect a worm as negligible. Thus, whenever a host in a monitored space became infected, it was detected and the number of infected hosts in the entire network was recorded.

We use three different monitored network sizes to see how fast we could detect random scan worms. We ran our experiment 15 times for each monitored network size and computed the average and variance. The result is shown in Table 5.4 (early warning time based on local victim information).

Table 5.4. Network simulator environments

Monitor Space	2^5	2^6	2^7
Avg Scan Rate	3.44	3.51	3.79
Avg Detection Time (%)	0.675	0.872	1.526
Variance	0.373	0.706	2.522

We can clearly see that using a smaller network, the detection time will be longer and the variance will be larger. Further, the detection time of the network simulation experiments matched the output of our analytical model when given identical input parameters.

Results and Analysis

With the information provided by the proposed detection algorithm, we can automatically take immediate and accurate responses that block victim port so as to effectively stop the propagation. The policy decision to block local ports can be accomplished entirely within the local network in realtime. This contrasts with global response strategies which require complex and time consuming coordination between CDC-like authority and Internet routers (Fig. 5.6).

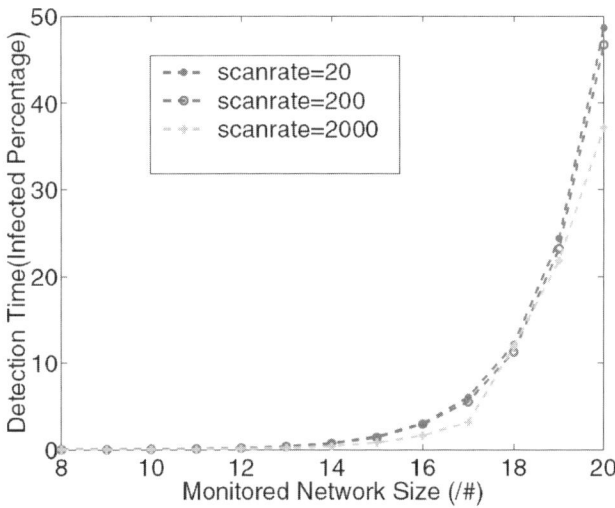

Fig. 5.6. Effect on different scan rates

Our local response is kind of Williamson's idea which limits the rate on the host by limiting the number of new connections [9]. Local response can be more effective, since local administrators know details about the victim machines and take more accurate action to block (not rate limit) the outgoing connections of victims (not all hosts) at that port (not all ports). This local response policy can be deployed on every host as [9]. We call this approach 'host level local response with random'. But it is probably more effective to only deploy on the edge router of LAN (network level local response) so that every LAN only need one position to deploy such a quarantine policy. We can also imagine a local response with preference because network level local response can only prevent victims from infecting outside network hosts but cannot prevent worm propagation within local network, which is worse with local preference scans.

For the network level response, the size of the local networks was set to be /27 in size. When a host is infected, all worm packets going out of the same /27 network were blocked, and vulnerable hosts inside could still be infected. The test was done for both uniform random and local preference scanning (scan policy: 0.25 in same /27, 0.25 in same /24, 0.25 in same /21, 0.25 in same /18). The result is shown in Figure 5.7 and 5.8.

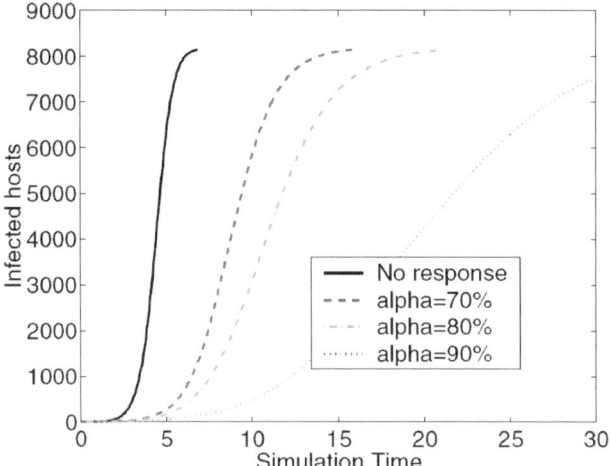

Fig. 5.7. Local response with random scan

As it can be seen from the Figure 5.7 and 5.8, the local response method was able to slow down the spread of the port scanning. With large numbers of hosts having a response capability, the worm propagation was slower. To compensate for the randomness of the network and target selection variation, each run was done several times to create an average propagation.

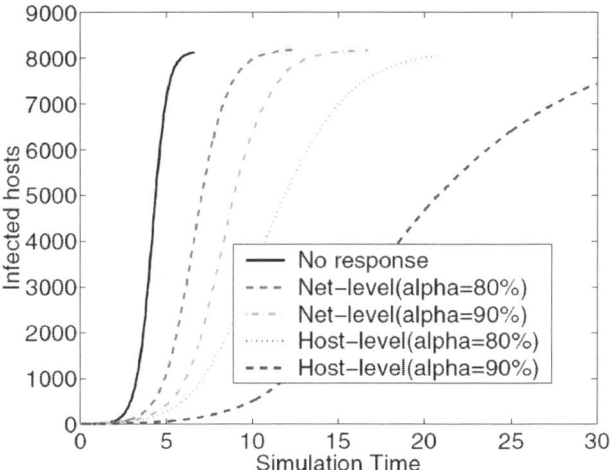

Fig. 5.8. Local response with preference scan

5.5 Conclusion

In this paper we have presented a detailed description of dendritic cells and the antigen presentation process from which an algorithm was abstracted. We have also discussed the relatively new Danger theory and given examples as to how Danger theory principles may be used in the field of AIS. We have also presented a worked example and prototype implementation based on this abstraction. The preliminary results are encouraging as both data orders produced low rates of false positive errors.

Many aspects of this algorithm remain unexplored such as the sensitivity of the parameters and scalability in terms of number of cells and number of input signals. For instance, we did not include any inflammatory cytokines in our examples due to data constraints. It would be interesting to explore their proposed amplifying effects on the other signals and on the behavior across a population of DCs. Additionally, an improved baseline for comparison must be developed in order to make serious claims regarding the effectiveness of the algorithm as an anomaly detector.

References

1. Forrest S. Hofmeyr S. Somayaji A. Computer immunology. *Communications of the ACM*, 40(10), (1997).
2. Shortman K. Caux C. Dendritic cell development: multiple pathways to nature's adjuvants. *Stem Cells*, 15:409–419, (1997).
3. Shortman K. Caux C. Dendritic cells and t lymphocytes: developmental and functional interactions. *Ciba Found Symp*, 204:130–141, (1997).

 4. Dasgupta D. An overview of artificial immune systems. *Artificial Immune Systems and Their Applications*, (1999).
 5. deCastro L. N. and Timmis J. Artificial immune systems a new computational intelligence approach. *Springer*, (2002).
 6. Satthaporn S. EreminO. Dendritic cells (i) : biological functions. *Journal of the Royal College of Surgeons of Edinburgh : Scientific Review*, 46(1), (2001).
 7. Steven Hofmeyr. An immunological model of distributed detection and its application to computer security. *PhD thesis*, (1999).
 8. Martien L. Kapsenberg. Dendritic-cell control of pathogen-driven t-cell polarization. *Nature Reviews in Immunology*, 3:984–993, December (2003).
 9. Williamson M. M. Throttling viruses. restricting propagation to defeat malicious mobile code. *Technical Report of HP Laboratories Bristol,*, HPL-2002-172, 2002.
10. Anderson C. Matzinger P. Danger: The view from the bottom of the cliff. *Seminars in Immunology*, 12(3):231–238, 2000.
11. Matzinger P. An innate sense of danger. *Seminars in Immunology*, 10(5):399–415, (1998).
12. Matzinger P. The danger model: A renewed sense of self. *Science*, 296:301–305, (2002).
13. G. J. et al. Randolph. Differentiation of monocytes into dendritic cells in a model of transendothelial trafficking see comments. *Science*, 282:480–483, (1998).
14. Percus Jerome K. Ora E. Alan S. Predicting the size of the t-cell receptor and antibody combining region from consideration of efficient self non-self discrimination. *Proceedings of the National Academy of Sciences of the United States of America*, 90:1691–1695, (1993).
15. Greensmith J. Aickelin U. Jamie T. Detecting danger. applying a novel immunological concept to intrusion detection systems. *6th International Conference in Design and Manufacture*, (2004).
16. Aickelin U. and Cayzer S. The danger theory and its application to artificial immune systems. *In proceedings of The First International Conference on Artificial Immune Systems (ICARIS 2002)*, pages 141–148, (2002).
17. Jamie T. Aickelin U. Towards a conceptual framework for innate immunity. *Proceedings of the 4th International Conference on Artificial Immune Systems*, (2004).
18. William R. Heath Jose A. Villadangos. No driving without a license. *Nature Immunology*, 6(2):125–126, (2005).
19. Jansen W. Intrusion detection with mobile agents. *Computer Communications*, 25(15), (2002).
20. Riley G. F. Sharif M. I. Lee W. Simulating internet worms. *Proceedings of IEEE/ACM MASCOTS 2004*, (2004).
21. Ki-Won Yeom and Ji-Hyung Park. An immune system inspired approach of collaborative intrusion detection system using mobile agents in wireless adhoc network. *Lecture Notes in Artificial Intelligence*, 3802:204–211, (2005).
22. Steinman R. Cohn Z. Identification of a novel cell type in peripheral lymphoid organs of mice. *The Journal of Experimental Medicine*, 137:1142–1162, (1973).

6

How to Efficiently Process Uncertainty within a Cyberinfrastructure without Sacrificing Privacy and Confidentiality

Luc Longpré and Vladik Kreinovich

Department of Computer Science, University of Texas at El Paso, 500 W. University, El Paso, TX 79968, USA, longpre@utep.edu, vladik@utep.edu

In this Chapter, we propose a simple solution to the problem of estimating uncertainty of the results of applying a black-box algorithm – without sacrificing privacy and confidentiality of the algorithm.

6.1 Cyberinfrastructure and Web Services

6.1.1 Practical Problem: Need to Combine Geographically Separate Computational Resources

In different knowledge domains in science and engineering, there is a large amount of data stored in different locations, and there are many software tools for processing this data, also implemented at different locations. Users may be interested in different information about this domain.

Sometimes, the information required by the user is already stored in *one of* the *databases*. For example, if we want to know the geological structure of a certain region in Texas, we can get his information from the geological map stored in Austin. In this case, all we need to do to get an appropriate response to the query is to get this data from the corresponding database.

In other cases, different pieces of the information requested by the user are *stored at different locations*. For example, if we are interested in the geological structure of the Rio Grande Region, then we need to combine data from the geological maps of Texas, New Mexico, and the Mexican state of Chihuahua. In such situations, a correct response to the user's query requires that we access these pieces of information from different databases located at different geographic notations.

In many other situations, the appropriate answer to the user's request requires that we not only collect the relevant data $x_1, \ldots, x_n$, but that we

L. Longpré and V. Kreinovich: *How to Efficiently Process Uncertainty within a Cyberinfrastructure without Sacrificing Privacy and Confidentiality*, Studies in Computational Intelligence (SCI) **57**, 155–173 (2007)

also use some *data processing* algorithms $f(x_1, \ldots, x_n)$ to process this data. For example, if we are interested in the large-scale geological structure of a geographical region, we may also use the gravity measurements from the gravity databases. For that, we need special algorithms to transform the values of gravity at different locations into a map that describes how the density changes with location. The corresponding data processing programs often require a lot of computational resources; as a result, many such programs reside on computers located at supercomputer centers, i.e., on computers which are physically separated from the places where the data is stored.

The need to combine computational resources (data and programs) located at different geographic locations seriously complicates research.

6.1.2 Centralization of Computational Resources – Traditional Approach to Combining Computational Resources; Its Advantages and Limitations

Traditionally, a widely used way to make these computational resources more accessible was to move all these resources to a *central location*. For example, in the geosciences, the US Geological Survey (USGS) was trying to become a central depository of all relevant geophysical data. However, this centralization requires a large amount of efforts: data are presented in different formats, the existing programs use specific formats, etc. To make the central data depository efficient, it is necessary:

- to reformat all the data,
- to rewrite all the data processing programs – so that they become fully compatible with the selected formats and with each other,
- etc.

The amount of work that is needed for this reformatting and rewriting is so large that none of these central depositories really succeeded in becoming an easy-to-use centralized database.

6.1.3 Cyberinfrastructure – A More Efficient Approach to Combining Computational Resources

Cyberinfrastructure technique is a new approach that provides the users with the efficient way to submit requests without worrying about the geographic locations of different computational resources – and at the same time avoid centralization with its excessive workloads. The main idea behind this approach is that *we keep all (or at least most) the computational resources*

- *at their current locations,*
- *in their current formats.*

To expedite the use of these resources:

- we supplement the local computational resources with the "metadata", i.e., with the information about the formats, algorithms, etc.,
- we "wrap up" the programs and databases with auxiliary programs that provide data compatibility into *web services*,

and, in general, we provide a cyberinfrastructure that uses the metadata to automatically combine different computational resources.

For example, if a user is interested in using the gravity data to uncover the geological structure of the Rio Grande region, then the system should automatically:

- get the gravity data from the UTEP and USGS gravity databases,
- convert them to a single format (if necessary),
- forward this data to the program located at San Diego Supercomputer Center, and
- move the results back to the user.

This example is exactly what we are designing under the NSF-sponsored Cyberinfrastructure for the Geosciences (GEON) project; see, e.g., [1–3, 8, 9, 27, 37, 45, 47, 48, 58, 62, 63]. This is similar to what other cyberinfrastructure projects are trying to achieve.

6.1.4 What Is Cyberinfrastructure: The Official NSF Definition

According to the final report of the National Science Foundation (NSF) Blue Ribbon Advisory Panel on Cyberinfrastructure, "a new age has dawned in scientific and engineering research, pushed by continuing progress in computing, information, and communication technology, and pulled by the expanding complexity, scope, and scale of today's challenges. The capacity of this technology has crossed thresholds that now make possible a comprehensive 'cyberinfrastructure' on which to build new types of scientific and engineering knowledge environments and organizations and to pursue research in new ways and with increased efficacy.

Such environments and organizations, enabled by cyberinfrastructure, are increasingly required to address national and global priorities, such as understanding global climate change, protecting our natural environment, applying genomics-proteomics to human health, maintaining national security, mastering the world of nanotechnology, and predicting and protecting against natural and human disasters, as well as to address some of our most fundamental intellectual questions such as the formation of the universe and the fundamental character of matter."

6.1.5 Web Services: What They Do – A Brief Summary

In different knowledge domains, there is a large amount of data stored in different locations; algorithms for processing this data are also implemented at different locations. Web services – and, more generally, cyberinfrastructure – provide the users with an efficient way to submit requests without worrying about the geographic locations of different computational resources (databases and programs) – and avoid centralization with its excessive workloads [21]. Web services enable the user to receive the desired data $x_1, \ldots, x_n$ and the results $y = f(x_1, \ldots, x_n)$ of processing this data.

6.2 Processing Uncertainty Within a Cyberinfrastructure

6.2.1 Formulation of the problem

The data x_i usually come from measurements or from experts. Measurements are never 100% accurate; as a result, the measured values $\widetilde{x}_i$ are, in general, somewhat different from the actual (unknown) values x_i of the corresponding quantities. Experts can also only provide us with approximate values of the desired quantities.

As a result of this measurement or expert uncertainty, the result $\widetilde{y} = f(\widetilde{x}_1, \ldots, \widetilde{x}_n)$ of data processing is, in general, different from the actual value $y = f(x_1, \ldots, x_n)$ of the desired quantity. It is desirable to gauge this difference. To do that, we must have some information about the errors of direct measurements.

Bounds on the measurement errors. What do we know about the errors $\Delta x_i \stackrel{\text{def}}{=} \widetilde{x}_i - x_i$ of direct measurements? First, the manufacturer of the measuring instrument must supply us with an upper bound Δ_i on the measurement error. If no such upper bound is supplied, this means that no accuracy is guaranteed, and the corresponding "measuring instrument" is practically useless. In this case, once we performed a measurement and got a measurement result $\widetilde{x}_i$, we know that the actual (unknown) value x_i of the measured quantity belongs to the interval $\mathbf{x}_i = [\underline{x}_i, \overline{x}_i]$, where $\underline{x}_i = \widetilde{x}_i - \Delta_i$ and $\overline{x}_i = \widetilde{x}_i + \Delta_i$.

Case of probabilistic uncertainty. In many practical situations, we not only know the interval $[-\Delta_i, \Delta_i]$ of possible values of the measurement error; we also know the probability of different values Δx_i within this interval. This knowledge underlies the traditional engineering approach to estimating the error of indirect measurement, in which we assume that we know the probability distributions for measurement errors Δx_i.

These probabilities are often described by a normal distribution, so in standard engineering textbook on measurement, it is usually assumed that the distribution of Δx_i is normal, with 0 average and known standard deviation σ_i; see, e.g. [20, 46].

In general, we can determine the desired probabilities of different values of Δx_i by comparing the results of measuring with this instrument with the results of measuring the same quantity by a standard (much more accurate) measuring instrument. Since the standard measuring instrument is much more accurate than the one use, the difference between these two measurement results is practically equal to the measurement error; thus, the empirical distribution of this difference is close to the desired probability distribution for measurement error.

Case of interval uncertainty. There are two cases, however, when in practice, we do not determine the probabilities:

- First is the case of cutting-edge measurements, e.g., measurements in fundamental science. When a Hubble telescope detects the light from a distant galaxy, there is no "standard" (much more accurate) telescope floating nearby that we can use to calibrate the Hubble: the Hubble telescope is the best we have.
- The second case is the case of measurements on the shop floor. In this case, in principle, every sensor can be thoroughly calibrated, but sensor calibration is so costly – usually costing ten times more than the sensor itself – that manufacturers rarely do it.

In both cases, we have no information about the probabilities of Δx_i; the only information we have is the upper bound on the measurement error.

In this case, after we performed a measurement and got a measurement result $\widetilde{x}_i$, the only information that we have about the actual value x_i of the measured quantity is that it belongs to the interval $\mathbf{x}_i = [\widetilde{x}_i - \Delta_i, \widetilde{x}_i + \Delta_i]$. In such situations, the only information that we have about the (unknown) actual value of $y = f(x_1, \ldots, x_n)$ is that y belongs to the range $\mathbf{y} = [\underline{y}, \overline{y}]$ of the function f over the box $\mathbf{x}_1 \times \ldots \times \mathbf{x}_n$:

$$\mathbf{y} = [\underline{y}, \overline{y}] = \{f(x_1, \ldots, x_n) \,|\, x_1 \in \mathbf{x}_1, \ldots, x_n \in \mathbf{x}_n\}.$$

The process of computing this interval range based on the input intervals $\mathbf{x}_i$ is called *interval computations*; see, e.g., [24, 26].

Case of fuzzy uncertainty. Often, knowledge comes in terms of uncertain expert estimates. In the fuzzy case, to describe this uncertainty, for each value of estimation error Δx_i, we describe the degree $\mu_i(\Delta x_i)$ to which this value is possible.

For each degree of certainty α, we can determine the set of values of Δx_i that are possible with at least this degree of certainty – the α-cut $\{x \,|\, \mu(x) \geq \alpha\}$ of the original fuzzy set. In most cases, this α-cut is an interval.

Vice versa, if we know α-cuts for every α, then, for each object x, we can determine the degree of possibility that x belongs to the original fuzzy set [7, 30, 38, 40, 41]. A fuzzy set can be thus viewed as a nested family of its α-cuts.

So, if instead of a (crisp) interval $\mathbf{x}_i$ of possible values of the measured quantity, we have a fuzzy set $\mu_i(x)$ of possible values, then we can view this information as a family of nested intervals $\mathbf{x}_i(\alpha)$ – α-cuts of the given fuzzy sets.

6.2.2 Description of uncertainty: general formulas

In this chapter, we will only consider a typical situation in which the direct measurements and/or expert estimates are accurate enough, so that the resulting approximation errors Δx_i are small, and terms which are quadratic (or of higher order) in Δx_i can be safely neglected. In such situations, the dependence of the desired value $y = f(x_1, \ldots, x_n) = f(\widetilde{x}_1 - \Delta x_1, \ldots, \widetilde{x}_n - \Delta x_n)$ on Δx_i can be safely assumed to be linear.

Comment. There are practical situations when the accuracy of the direct measurements is not high enough, and hence, quadratic terms cannot be safely neglected (see, e.g., [26] and references therein). In this case, the problem of error estimation for indirect measurements becomes computationally difficult (NP-hard) even when the function $f(x_1, \ldots, x_n)$ is quadratic [34,60]. However, in most real-life situations, the possibility to ignore quadratic terms is a reasonable assumption, because, e.g., for an error of 1% its square is a negligible 0.01%.

When approximation errors are small, we can simplify the expression for $\Delta y = \widetilde{y} - y = f(\widetilde{x}_1, \ldots, \widetilde{x}_n) - f(x_1, \ldots, x_n)$ if we expand the function f in Taylor series around the point $(\widetilde{x}_1, \ldots, \widetilde{x}_n)$ and restrict ourselves only to linear terms in this expansion. As a result, we get the expression

$$\Delta y = c_1 \cdot \Delta x_1 + \ldots + c_n \cdot \Delta x_n, \tag{6.1}$$

where by c_i, we denoted the value of the partial derivative $\partial f / \partial x_i$ at the point $(\widetilde{x}_1, \ldots, \widetilde{x}_n)$:

$$c_i = \frac{\partial f}{\partial x_i}_{|(\widetilde{x}_1, \ldots, \widetilde{x}_n)}. \tag{6.2}$$

Case of probabilistic uncertainty. In the statistical setting, the desired measurement error Δy is a linear combination of independent Gaussian variables Δx_i. Therefore, Δy is also normally distributed, with 0 average and the standard deviation

$$\sigma = \sqrt{c_1^2 \cdot \sigma_1^2 + \ldots + c_n^2 \cdot \sigma_n^2}. \tag{6.3}$$

Comment. A similar formula holds if we *do not* assume that Δx_i are normally distributed: it is sufficient to assume that they are independent variables with 0 average and known standard deviations σ_i.

Case of interval uncertainty. In the interval setting, we do not know the probability of different errors Δx_i; instead, we only know that $|\Delta x_i| \leq \Delta_i$. In this case, the sum (6.1) attains its largest possible value if each term $c_i \cdot \Delta x_i$ in this sum attains the largest possible value:

- If $c_i \geq 0$, then this term is a monotonically non-decreasing function of Δx_i, so it attains its largest value at the largest possible value $\Delta x_i = \Delta_i$; the corresponding largest value of this term is $c_i \cdot \Delta_i$.

- If $c_i < 0$, then this term is a decreasing function of Δx_i, so it attains its largest value at the smallest possible value $\Delta x_i = -\Delta_i$; the corresponding largest value of this term is $-c_i \cdot \Delta_i = |c_i| \cdot \Delta_i$.

In both cases, the largest possible value of this term is $|c_i| \cdot \Delta_i$, so, the largest possible value of the sum Δy is

$$\Delta = |c_1| \cdot \Delta_1 + \ldots + |c_n| \cdot \Delta_n. \tag{6.4}$$

Similarly, the smallest possible value of Δy is $-\Delta$.

Hence, the interval of possible values of Δy is $[-\Delta, \Delta]$, with Δ defined by the formula (6.4).

Case of fuzzy uncertainty. We have already mentioned that if instead of a (crisp) interval $\mathbf{x}_i$ of possible values of the measured quantity, we have a fuzzy set $\mu_i(x)$ of possible values, then we can view this information as a family of nested intervals $\mathbf{x}_i(\alpha)$ – α-cuts of the given fuzzy sets.

Our objective is then to compute the fuzzy number corresponding to this the desired value $y = f(x_1, \ldots, x_n)$. In this case, for each level α, to compute the α-cut of this fuzzy number, we can apply interval computations to the α-cuts $\mathbf{x}_i(\alpha)$ of the corresponding fuzzy sets. The resulting nested intervals form the desired fuzzy set for y.

So, e.g., if we want to describe 10 different levels of uncertainty, then we must solve 10 interval computation problems – i.e., apply the formula (6.4) 10 times.

In many practical situations, there is no need to perform 10 computations. For example, it is often reasonable to assume that all the membership functions $\mu_i(x_i)$ have the same shape and only differ by a scaling parameter, i.e., all have the form $\mu_i(\Delta x_i) = \mu_0(\Delta x_i/\Delta_i)$ for some fixed function $\mu_0(x)$ (e.g., triangular or Gaussian). In this case, as it is well known [14–16, 25], the membership function for Δy has a similar form $\mu_0(\Delta y/\Delta)$, where Δ is determined by the formula (6.4).

Comment. These formulas correspond to the case when in the *extension principle* that describe how the uncertainty in Δx_i transforms into the uncertainty in Δy, we interpret "and" as min, i.e., if we consider

$$\mu(\Delta y) = \max_{\Delta x_1, \ldots, \Delta x_n} \min(\mu_1(\Delta x_1), \ldots, \mu_n(\Delta x_n)), \tag{6.5}$$

where maximum is taken over all the values $\Delta x_1, \ldots, \Delta x_n$ for which the expression (6.1) for Δy leads to the given value of Δy. In general, we can use a different t-norm to combine the values $\mu_i(\Delta x_i)$. For example, we may use the product and describe the resulting membership function as

$$\mu(\Delta y) = \max_{\Delta x_1, \ldots, \Delta x_n} \mu_1(\Delta x_1) \cdot \ldots \cdot \mu_n(\Delta x_n). \tag{6.6}$$

In this case, if we assume that all the membership functions $\mu_i(\Delta x_i)$ are Gaussian, i.e., have the form $\mu_i(\Delta x_i) = \mu_0(\Delta x_i/\sigma_i)$, where $\mu_0(z) = \exp(-z^2)$, then the resulting membership function for Δy is also Gaussian $\mu(\Delta y) = \mu_0(\Delta y/\sigma)$, where σ is determined by the formula (6.3); see, e.g., [14–16, 25].

6.2.3 Error Estimation for the Results of Data Processing: A Precise Computational Formulation of the Problem

As a result of the above analysis, we get the following explicit formulation of the problem: given a function $f(x_1, \ldots, x_n)$, n numbers $\tilde{x}_1, \ldots, \tilde{x}_n$, and n positive numbers $\sigma_1, \ldots, \sigma_n$ (or $\Delta_1, \ldots, \Delta_n$), compute the corresponding expression (6.3) or (6.4).

Let us describe how this problem is solved now.

6.2.4 How This Problem Is Solved Now

Textbook case: the function f is given by its analytical expression. If the function f is given by its analytical expression, then we can simply explicitly differentiate it, and get an explicit expression for (6.3) and (6.4). This is the case which is typically analyzed in textbooks on measurement theory; see, e.g., [20, 46].

A more complex case: automatic differentiation. In many practical cases, we do not have an explicit analytical expression, we only have an *algorithm* for computing the function $f(x_1, \ldots, x_n)$, an algorithm which is too complicated to be expressed as an analytical expression.

When this algorithm is presented in one of the standard programming languages such as Fortran or C, we can let the compute perform an explicit differentiation; for that, we can use one of the existing automatic differentiation tools (see, e.g., [6, 22]). These tools analyze the code of the program for computing $f(x_1, \ldots, x_n)$ and, as they perform their analysis, they produce the "differentiation code", i.e., a program that computes the partial derivatives c_i.

Once we know an algorithm that computes f in time T, automatic differentiation (AD) enables us to compute all partial derivatives in time $\leq 3T$, hence we can compute σ or Δ in time $O(T + n)$.

6.3 Need for Privacy Makes the Problem More Complex

Privacy situation: description. In cyberinfrastructure, the owners of the program f may not want to disclose its code; instead, they may only allow to use f as a black box.

Real world example. For example, to find places where oil can be found, it is important to know the structure of the Earth crust at different locations. One of the main techniques for determining this structure is the use of seismic data: we use the measured travel-times of the natural (or artificial) seismic signals as they go from their source to their on-surface destination, and then we solve the corresponding *inverse problem* to find the Earth structure; see, e.g., [2–5, 11, 13, 23, 44, 64].

Because of the importance of the seismic inverse problem, oil companies are heavily investing in developing algorithms for solving such problems. This is a very competitive area of research, so the algorithms are kept confidential. Since most major algorithms have similar efficiency and accuracy, a company is willing to allow other users – researchers or competitors from other companies – to actually use their code. However, companies are very reluctant to let the users have actual access to their code. The reason for this reluctance is that different companies have achieved the similar level of efficiency and accuracy by using different ideas and programming improvements. As a result, a company that unilaterally discloses its code would thus let competitors use its ideas (and thus, improve their code) without getting any benefits in return.

Thus, the companies are allowing to use their code, but only as a black box.

What was known before. The problem of preserving the code's privacy and confidentiality is a particular case of a general problem of privacy preservation in data processing; see, e.g., [10, 12, 17–19, 28, 29, 35, 39, 42, 43, 49–57, 61]. At present, privacy-preserving algorithms are mainly concerned with privacy of *data*; privacy of *code* is a problem for which few results are known.

Related difficulty. If we do not know the code of f, then we cannot apply AD to compute all n partial derivatives $c_i = \dfrac{\partial f}{\partial x_i}$.

A straightforward method of solving this problem: numerical differentiation. The most straightforward algorithm for solving this problem is to compute the derivatives c_i one-by-one, and then use the corresponding formula (6.3) or (6.4) to compute the desired σ. To compute the i-th partial derivative, we change the i-th input x_i to $\tilde{x}_i + h_i$ for some h_i, and leave other inputs unchanged, i.e., we take $\delta_i = h_i$ for this i and $\delta_j = 0$ for all $j \neq i$. Then, we estimate c_i as

$$c_i = \frac{f\left(\tilde{x}_1, \ldots, \tilde{x}_{i-1}, \tilde{x}_i + h_i, \tilde{x}_{i+1}, \ldots, \tilde{x}_n\right) - \tilde{y}}{h_i}. \tag{6.7}$$

This algorithm is called *numerical differentiation*.

We want the change h_i to be small (so that quadratic terms can be neglected); we already know that changes of the order σ_i are small. So, it is natural to take $h_i = \sigma_i$ (or, correspondingly, $h_i = \Delta_i$). In other words, to compute c_i, we use the following values: $\delta_1 = \ldots = \delta_{i-1} = 0$, $\delta_i = \sigma_i$ (or $\delta_i = \Delta_i$), $\delta_{i+1} = \ldots = \delta_n = 0$.

Problem: sometimes, numerical differentiation takes too long. Very often, the program f requires a reasonable time to compute (e.g., in the geological applications, computing f may involve solving an inverse problem). In this case, applying the function f is the most time-consuming part of this algorithm. So, the total time that it takes us to compute σ or Δ is (approximately) equal to the running time T for the program f multiplied by the number of times N_f that we call the program f.

For numerical differentiation, $N_f = n$ (we call f n times to compute n partial derivatives). Hence, if the program f takes a long time to compute, and n is huge, then the resulting time $T \cdot n$ (which is $\gg T + n$) may be too long. For example, if we are determining some parameters of an oil well from the geophysical measurements, we may get n in the thousands, and T in minutes. In this case, $T \cdot n$ may take several weeks. This may be OK for a single measurement, but too long if we want more on-line results.

6.4 Solution for Statistical Setting: Monte-Carlo Simulations

Monte-Carlo simulations: main idea. In the statistical setting, we can use straightforward (Monte-Carlo) simulation, and drastically save the computation time. In this approach, we use a computer-based random number generator to simulate the normally distributed error. A standard normal random number generator usually produces a normal distribution with 0 average and standard deviation 1. So, to simulate a distribution with a standard deviation σ_i, we multiply the result α_i of the standard random number generator by σ_i. In other words, we take $\delta_i = \sigma_i \cdot \alpha_i$.

As a result of N Monte-Carlo simulations, we get N values $c^{(1)} = \mathbf{c} \cdot \boldsymbol{\delta}^{(1)}, \ldots, c^{(N)} = \mathbf{c} \cdot \boldsymbol{\delta}^{(N)}$ which are normally distributed with the desired standard deviation σ. So, we can determine σ by using the standard statistical estimate

$$\sigma = \sqrt{\frac{1}{N-1} \cdot \sum_{k=1}^{N} \left(c^{(k)}\right)^2}. \tag{6.8}$$

Computation time required for Monte-Carlo simulation. The relative error of the above statistical estimate depends only on N (as $\approx 1/\sqrt{N}$), and not on the number of variables n. Therefore, the number N_f of calls to f that is needed to achieve a given accuracy does not depend on the number of variables at all.

The error of the above algorithm is asymptotically normally distributed, with a standard deviation $\sigma_e \sim \sigma/\sqrt{2N}$. Thus, if we use a "two sigma" bound, we conclude that with probability 95%, this algorithm leads to an estimate for σ which differs from the actual value of σ by $\leq 2\sigma_e = 2\sigma/\sqrt{2N}$.

This is an error with which we estimate the error of indirect measurement; we do not need too much accuracy in this estimation, because, e.g., in real life, we say that an error is $\pm 10\%$ or $\pm 20\%$, but *not* that the error is, say, $\pm 11.8\%$. Therefore, in estimating the error of indirect measurements, it is sufficient to estimate the characteristics of this error with a relative accuracy of, say, 20%.

For the above "two sigma" estimate, this means that we need to select the smallest N for which $2\sigma_e = 2\sigma/\sqrt{2N} \leq 0.2 \cdot \sigma$, i.e., to select $N_f = N = 50$.

In many practical situations, it is sufficient to have a standard deviation of 20% (i.e., to have a "two sigma" guarantee of 40%). In this case, we need only $N = 13$ calls to f.

On the other hand, if we want to guarantee 20% accuracy in 99.9% cases, which correspond to "three sigma", we must use N for which $3\sigma_e = 3 \cdot \sigma/\sqrt{2N} \leq 0.2 \cdot \sigma$, i.e., we must select $N_f = N = 113$, etc.

For $n \approx 10^3$, all these values of N_f are much smaller than $N_f = n$ required for numerical differentiation.

So, if we have to choose between the (deterministic) numerical differentiation and the randomized Monte-Carlo algorithm, we must select:

- a deterministic algorithm when the number of variables n satisfies the inequality $n \leq N_0$ (where $N_0 \approx 50$), and
- a randomized method if $n \geq N_0$.

Additional advantage: parallelization. In Monte-Carlo algorithm, we need 50 calls to f. If each call requires a minute, the resulting time takes about an hour, which may be too long for on-line results. Fortunately, different calls to the function f are independent on each other, so we can run all the simulations in parallel.

The more processors we have, the less time the resulting computation will take. If we have as many processors as the required number of calls, then the time needed to estimate the error of indirect measurement becomes equal to the time of a single call, i.e., to the time necessary to compute the result $\widetilde{y}$ of this indirect measurement. Thus, if we have enough processors working in parallel, we can compute the result of the indirect measurement *and* estimate its error during the same time that it normally takes just to compute the result.

In particular, if the result $\widetilde{y}$ of indirect measurement can be computed in real time, we can estimate the error of this result in real time as well.

6.5 Solution for Interval and Fuzzy Setting: New Method Based on Cauchy Distribution

Can we use a similar idea in the interval setting? Since Monte-Carlo simulation speeds up computations, it is desirable to use a similar technique in interval setting as well.

There is a problem here. In the interval setting, we do not know the exact distribution, we may have different probability distributions – as long as they are located within the corresponding intervals. If we only use one of these distributions for simulations, there is no guarantee that the results will be valid for other distributions as well.

In principle, we could repeat simulations for several different distributions, but this repetition would drastically increase the simulation time and thus, eliminate the advantages of simulation as opposed to numerical differentiation.

Yes, we can. Luckily, there is a mathematical trick that enables us to use Monte-Carlo simulation in interval setting as well. This trick is based on using *Cauchy distribution* – i.e., probability distributions with the probability density

$$\rho(z) = \frac{\Delta}{\pi \cdot (z^2 + \Delta^2)};$$ (6.9)

the value Δ is called the *scale parameter* of this distribution, or simply a *parameter*, for short.

Cauchy distribution has the following property that we will use: if $z_1, \ldots, z_n$ are independent random variables, and each of z_i is distributed according to the Cauchy law with parameter Δ_i, then their linear combination $z = c_1 \cdot z_1 + \ldots + c_n \cdot z_n$ is also distributed according to a Cauchy law, with a scale parameter $\Delta = |c_1| \cdot \Delta_1 + \ldots + |c_n| \cdot \Delta_n$.

Therefore, if we take random variables δ_i which are Cauchy distributed with parameters Δ_i, then the value

$$c = f(\widetilde{x}_1 + \delta_1, \ldots, \widetilde{x}_n + \delta_n) - f(\widetilde{x}_1, \ldots, \widetilde{x}_n) = c_1 \cdot \delta_1 + \ldots + c_n \cdot \delta_n \quad (6.10)$$

is Cauchy distributed with the desired parameter (6.4). So, repeating this experiment N times, we get N values $c^{(1)}, \ldots, c^{(N)}$ which are Cauchy distributed with the unknown parameter, and from them we can estimate Δ.

The bigger N, the better estimates we get.

There are two questions to be solved:

- how to simulate the Cauchy distribution;
- how to estimate the parameter Δ of this distribution from a finite sample.

Simulation can be based on the functional transformation of uniformly distributed sample values:

$$\delta_i = \Delta_i \cdot \tan(\pi \cdot (r_i - 0.5)),$$ (6.11)

where r_i is uniformly distributed on the interval $[0, 1]$.

In order to estimate σ, we can apply the Maximum Likelihood Method $\rho(d^1) \cdot \rho(d^2) \cdot \ldots \cdot \rho(d^n) \to \max$, where $\rho(z)$ is a Cauchy distribution density with the unknown Δ. When we substitute the above-given formula for $\rho(z)$ and equate the derivative of the product with respect to Δ to 0 (since it is a maximum), we get an equation

$$\frac{1}{1 + \left(\frac{c^{(1)}}{\Delta}\right)^2} + \ldots + \frac{1}{1 + \left(\frac{c^{(N)}}{\Delta}\right)^2} = \frac{N}{2}.$$ (6.12)

The left-hand side of (6.12) is an increasing function that is equal to $0 (< N/2)$ for $\Delta = 0$ and $> N/2$ for $\Delta = \max |c^{(k)}|$; therefore the solution to the equation (6.12) can be found by applying a bisection method to the interval $\left[0, \max |c^{(k)}|\right]$.

It is important to mention that we assumed that the function f is reasonably linear within the box

$$[\widetilde{x}_1 - \Delta_1, \widetilde{x}_1 + \Delta_1] \times \ldots \times [\widetilde{x}_n - \Delta_n, \widetilde{x}_n + \Delta_n]. \tag{6.13}$$

However, the simulated values δ_i may be outside the box. When we get such values, we do not use the function f for them, we use a normalized function that is equal to f within the box, and that is extended linearly for all other values (we will see, in the description of an algorithm, how this is done).

As a result, we arrive at the following algorithm (described, for a somewhat different problem, in [32, 33, 36, 59]):

Algorithm.

- Apply f to the results of direct measurements: $\widetilde{y} := f(\widetilde{x}_1, \ldots, \widetilde{x}_n)$;
- For $k = 1, 2, \ldots, N$, repeat the following:
  - use the standard random number generator to compute n numbers $r_i^{(k)}$, $i = 1, 2, \ldots, n$, that are uniformly distributed on the interval $[0, 1]$;
  - compute Cauchy distributed values $c_i^{(k)} := \tan(\pi \cdot (r_i^{(k)} - 0.5))$;
  - compute the largest value of $|c_i^{(k)}|$ so that we will be able to normalize the simulated measurement errors and apply f to the values that are within the box of possible values: $K := \max_i |c_i^{(k)}|$;
  - compute the simulated measurement errors $\delta_i^{(k)} := \Delta_i \cdot c_i^{(k)}/K$;
  - compute the simulated measurement results $x_i^{(k)} := \widetilde{x}_i + \delta_i^{(k)}$;
  - apply the program f to the simulated measurement results and compute the simulated error of the indirect measurement:

$$c^{(k)} := K \cdot \left(f\left(x_1^{(k)}, \ldots, x_n^{(k)}\right) - \widetilde{y} \right);$$

- Compute Δ by applying the bisection method to solve the equation (6.12).

When is this randomized algorithm better than deterministic numerical differentiation? To determine the parameter Δ, we use the maximum likelihood method. It is known that the error of this method is asymptotically normally distributed, with 0 average and standard deviation $1/\sqrt{N \cdot I}$, where I is Fisher's information:

$$I = \int_{-\infty}^{\infty} \frac{1}{\rho} \cdot \left(\frac{\partial \rho}{\partial \Delta}\right)^2 \, dz.$$

For Cauchy probability density $\rho(z)$, we have $I = 1/(2\Delta^2)$, so the error of the above randomized algorithm is asymptotically normally distributed, with a standard deviation $\sigma_e \sim \Delta \cdot \sqrt{2/N}$. Thus, if we use a "two sigma" bound, we conclude that with probability 95%, this algorithm leads to an estimate for Δ which differs from the actual value of Δ by $\leq 2\sigma_e = 2\Delta \cdot \sqrt{2/N}$. So, if we want to achieve a 20% accuracy in the error estimation, we must use the smallest N for which $2\sigma_e = 2\Delta \cdot \sqrt{2/N} \le 0.2 \cdot \Delta$, i.e., to select $N_f = N = 200$.

When it is sufficient to have a standard deviation of 20% (i.e., to have a "two sigma" guarantee of 40%), we need only $N = 50$ calls to f. For $n \approx 10^3$, both values N_f are much smaller than $N_f = n$ required for numerical differentiation.

So, if we have to choose between the (deterministic) numerical differentiation and the randomized Monte-Carlo algorithm, we must select:

- a deterministic algorithm when the number of variables n satisfies the inequality $n \le N_0$ (where $N_0 \approx 200$), and
- a randomized algorithm if $n \ge N_0$.

Comment. If we use fewer than N_0 simulations, then we still get an approximate value of the range, but with worse accuracy – and the accuracy can be easily computed by using the above formulas.

This algorithm is naturally parallelizable. Similarly to the Monte-Carlo algorithm for statistical setting, we can run all N simulations in parallel and thus, speed up the computations.

Conclusion. When we know the code for f, then we can use AD and compute Δ and σ in time $O(T + n)$. If the owner of the program f only allows to use it as a black box, then we cannot use AD any more. In principle, we can compute each of n derivatives $\partial f / \partial x_i$ by numerical differentiation, but this would require computation time $T \cdot n \gg T + n$.

For probabilistic uncertainty, one can use Monte-Carlo simulations and compute σ in time $O(T) \ll T \cdot n$. We have shown that for interval uncertainty, we can also compute Δ in time $O(T)$ by using an artificial Monte-Carlo simulations in which each Δx_i is Cauchy distributed with parameter Δ_i – then simulated Δy is Cauchy distributed with the desired parameter Δ.

Remark: the problem of non-linearity. In the above text, we assumed that the intervals $\mathbf{x}_i$ are narrow. In this case, terms quadratic in Δx_i are negligible, and so, we can safely assume that the desired function $f(x_1, \ldots, x_n)$ is linear on the box

$$\mathbf{x}_1 \times \ldots \times \mathbf{x}_n.$$

In practice, some intervals $\mathbf{x}_i$ may be wide, so even when restricted to the box, the function $f(x_1, \ldots, x_n)$ is non-linear. What can we do in this case?

Usually, experts (e.g., designers of the corresponding technical system) know for which variables x_i, the dependence is non-linear. For each of these variables, we can *bisect* the corresponding interval $[\underline{x}_i, \overline{x}_i]$ into two smaller subintervals – for which the dependence is approximately linear. Then, we estimate the range of the function f separately on each of the resulting subboxes, and take the union of these two ranges as the range over the entire box.

If one bisection is not enough and the dependence of f on x_i is non-linear over one or several subboxes, we can bisect these boxes again, etc.

This bisection idea has been successfully used in interval computations; see, e.g., [24, 26].

6.6 Summary

In different knowledge domains, there is a large amount of data stored in different locations; algorithms for processing this data are also implemented at different locations. Web services – and, more generally, cyberinfrastructure – provide the users with an efficient way to submit requests without worrying about the geographic locations of different computational resources (databases and programs) – and avoid centralization with its excessive workloads [21]. Web services enable the user to receive the desired data $x_1, \ldots, x_n$ and the results $y = f(x_1, \ldots, x_n)$ of processing this data.

The data x_i usually come from measurements or from experts. Measurements are never 100% accurate; as a result, the measured values $\widetilde{x}_i$ are, in general, somewhat different from the actual (unknown) values x_i of the corresponding quantities. Experts can also only provide us with approximate values of the desired quantities.

As a result of this measurement or expert uncertainty, the result $\widetilde{y} = f(\widetilde{x}_1, \ldots, \widetilde{x}_n)$ of data processing is, in general, different from the actual value $y = f(x_1, \ldots, x_n)$ of the desired quantity. It is desirable to gauge this difference.

Traditional methods for estimating the resulting uncertainty in y are based on the assumption that we know the code of the function f. In cyberinfrastructure, owners of the program f may not want to disclose its code; instead, they may only allow to use f as a black box. In this case, traditional techniques are not applicable.

There exist techniques for processing uncertainty under such a "black-box" situation, but these techniques require much longer computation time. In this chapter, we describe new Monte-Carlo-type techniques that process uncertainty in such privacy-protecting black-box situations and that require the same amount of computation time as the traditional non-privacy-protecting techniques.

Acknowledgments

This work was supported in part by NASA under cooperative agreement NCC5-209, NSF grants EAR-0225670 and DMS-0532645, Star Award from the University of Texas System, and Texas Department of Transportation grant No. 0-5453. The authors are thankful to the anonymous referees for valuable suggestions.

References

1. Aguiar MS, Dimuro GP, Costa ACR, Silva RKS, Costa FA, Kreinovich V (2004) The multi-layered interval categorizer tesselation-based model, In: Iochpe C, Câmara G (eds), IFIP WG2.6 Proceedings of the 6th Brazilian Symposium on Geoinformatics Geoinfo'2004, Campos do Jordão, Brazil, November 22–24, 2004, pp. 437–454. ISBN 3901882200
2. Aldouri R, Keller GR, Gates A, Rasillo J, Salayandia L, Kreinovich V, Seeley J, Taylor P, Holloway S (2004) GEON: Geophysical data add the 3rd dimension in geospatial studies. In: Proceedings of the ESRI International User Conference 2004, San Diego, California, August 9–13, 2004, Paper 1898
3. Averill MG, Miller KC, Keller GR, Kreinovich V, Araiza R, Starks SA (2005) Using expert knowledge in solving the seismic inverse problem, In: Proceedings of the 24nd International Conference of the North American Fuzzy Information Processing Society NAFIPS'2005, Ann Arbor, Michigan, June 22–25, 2005, pp. 310–314
4. Averill MG, Miller KC, Keller GR, Kreinovich V, Araiza R, Starks SA (2007) Using Expert Knowledge in Solving the Seismic Inverse Problem. International Journal of Approximate Reasoning, to appear
5. Bardossy G, Fodor J (2004) Evaluation of uncertainties and risks in geology. Springer Verlag, Berlin
6. Berz M, Bischof C, Corliss G, Griewank A (1996), Computational differentiation: techniques, applications, and tools. SIAM, Philadelphia
7. Bojadziev G, Bojadziev M (1995) Fuzzy sets, fuzzy logic, applications. World Scientific, Singapore
8. Ceberio M, Ferson S, Kreinovich V, Chopra S, Xiang G, Murguia A, Santillan J (2006) How to take into account dependence between the inputs: from interval computations to constraint-related set computations, with potential applications to nuclear safety, bio- and geosciences. In: Proceedings of the Second International Workshop on Reliable Engineering Computing, Savannah, Georgia, February 22–24, 2006, pp. 127–154
9. Ceberio M, Kreinovich V, Chopra S, Ludäscher B (2005) Taylor model-type techniques for handling uncertainty in expert systems, with potential applications to geoinformatics. In: Proceedings of the 17th World Congress of the International Association for Mathematics and Computers in Simulation IMACS'2005, Paris, France, July 11–15, 2005
10. Dalenius T (1986) Finding a needle in a haystack – or identifying anonymous census record. Journal of Official Statistics 2(3):329–336
11. Demicco R, Klir G, eds (2003) Fuzzy logic in geology. Academic Press
12. Denning DERD (1982) Cryptography and data security. Addison-Wesley, Reading, Massachusetts
13. Doser DI, Crain KD, Baker MR, Kreinovich V, Gerstenberger MC (1998) Estimating uncertainties for geophysical tomography. Reliable Computing 4(3):241–268
14. Dubois D, Prade H (1978) Operations on fuzzy numbers. International Journal of Systems Science 9:613–626
15. Dubois D, Prade H (1979) Fuzzy real algebra: some results. Fuzzy Sets and Systems 2:327–348
16. Dubois D, Prade H (1980) Fuzzy sets and systems: theory and applications. Academic Press, New York, London

17. Duncan G, Lambert D (1987) The risk of disclosure for microdata. In: *Proc. of the Bureau of the Census Third Annual Research Conference*, Bureau of the Census, Washington, DC, 263–274

18. Duncan G, Mukherjee S (1991) Microdata disclosure limitation in statistical databases: query size and random sample query control. In: Prof. 1991 IEEE Symposium on Research in Security and Privacy, Oakland, CA, May 20–22, 1991

19. Fellegi I (1972) On the question of statistical confidentiality. Journal of the American Statistical Association 7–18

20. Fuller WA (1987) Measurement error models. J. Wiley & Sons, New York

21. Gates A, Kreinovich V, Longpré L, Pinheiro da Silva P, Keller GR (2006) Towards secure cyberinfrastructure for sharing border information. In: Proceedings of the Lineae Terrarum: International Border Conference, El Paso, Las Cruces, and Cd. Juárez, March 27–30, 2006

22. Griewank A (2000) Evaluating derivatives: principles and techniques of algorithmic differentiation. SIAM, Philadelphia

23. Hole JA (1992) Nonlinear high-resolution three-dimensional seismic travel time tomography. J. Geophysical Research 97(B5):6553–6562.

24. Jaulin L, Kieffer M, Didrit O, Walter E (2001) Applied interval analysis. Springer Verlag, London

25. Kauffman A, Gupta MM (1985) Introduction to fuzzy arithmetic: theory and applications. Van Nostrand, New York

26. Kearfott RB, Kreinovich V, eds (1996) Applications of interval computations. Kluwer, Dordrecht

27. Keller GR, Hildenbrand TG, Kucks R, Webring M, Briesacher A, Rujawitz K, Hittleman AM, Roman DJ, Winester D, Aldouri R, Seeley J, Rasillo J, Torres T, Hinze WJ, Gates A, Kreinovich V, Salayandia L (2006) A community effort to construct a gravity database for the United States and an associated Web portal. In: Sinha AK (ed), Geoinformatics: Data to Knowledge, Geological Society of America Publ., Boulder, Colorado, pp. 21–34

28. Kim J (1986) A method for limiting disclosure of microdata based on random noise and transformation. In: Proceedings of the Section on Survey Research Methods of the American Statistical Association 370–374

29. Kirkendall N et al. (1994) Report on Statistical Disclosure Limitations Methodology. Office of Management and Budget, Washington, DC, Statistical Policy Working Paper No. 22

30. Klir G, Yuan B (1995) Fuzzy sets and fuzzy logic. Prentice Hall, New Jersey

31. Kreinovich V, Beck J, Ferregut C, Sanchez A, Keller GR, Averill M, Starks SA (2007) Monte-Carlo-type techniques for processing interval uncertainty, and their potential engineering applications. Reliable Computing 13(1):25–69.

32. Kreinovich V, Bernat A, Villa E, Mariscal Y (1991) Parallel computers estimate errors caused by imprecise data. Interval Computations (2):21–46

33. Kreinovich V, Ferson S (2004) A new cauchy-based black-box technique for uncertainty in risk analysis. Reliability Engineering and Systems Safety 85 (1–3):267–279

34. Kreinovich V, Lakeyev A, Rohn J, Kahl P (1998) Computational complexity and feasibility of data processing and interval computations. Kluwer, Dordrecht

35. Kreinovich V, Longpré L, Starks SA, Xiang G, Beck J, Kandathi K, Nayak A, Ferson S, Hajagos J (2007) Interval versions of statistical techniques, with

applications to environmental analysis, bioinformatics, and privacy in statistical databases. Journal of Computational and Applied Mathematics 199(2):418–423

36. Kreinovich V, Pavlovich MI (1985) Error estimate of the result of indirect measurements by using a calculational experiment. Measurement Techniques 28(3):201–205

37. Longpré L, Kreinovich V, Freudenthal E, Ceberio M, Modave F, Baijal N, Chen W, Chirayath V, Xiang G, Vargas JI (2005) Privacy: protecting, processing, and measuring loss. In: Abstracts of the 2005 South Central Information Security Symposium SCISS'05, Austin, Texas, April 30, 2005, p. 2

38. Moore RE, Lodwick WA (2003) Interval analysis and fuzzy set theory. Fuzzy Sets and Systems 135(1):5–9

39. Morgenstern M (1987) Security and inference in multilevel database and knowledge base systems. In: Proc. of the ACM SIGMOD Conference 357–373

40. Nguyen HT, Kreinovich V (1996) Nested intervals and sets: concepts, relations to fuzzy sets, and applications, In: Kearfott RB, Kreinovich V (eds) Applications of interval computations. Kluwer, Dordrecht, pp. 245–290.

41. Nguyen HT, Walker EA (2005) First course in fuzzy logic. CRC Press, Boca Raton, Florida

42. Office of Technology Assessment (1993) Protecting privacy in computerized medical information. US Government Printing Office, Washington, DC

43. Palley M, Siminoff J (1986) Regression methodology based disclosure of a statistical database. In: Proceedings of the Section on Survey Research Methods of the American Statistical Association 382–387

44. Parker RL (1994) Geophysical inverse theory. Princeton University Press, Princeton, New Jersey

45. Platon E, Tupelly K, Kreinovich V, Starks SA, Villaverde K (2005) Exact bounds for interval and fuzzy functions under monotonicity constraints, with potential applications to biostratigraphy. In: Proceedings of the 2005 IEEE International Conference on Fuzzy Systems FUZZ-IEEE'2005, Reno, Nevada, May 22–25, 2005, pp. 891–896

46. Rabinovich S (2005) Measurement errors and uncertainties: theory and practice. American Institute of Physics, New York

47. Schiek CG, Araiza R, Hurtado JM, Velasco AA, Kreinovich V, Sinyansky V (2006) Images with uncertainty: efficient algorithms for shift, rotation, scaling, and registration, and their applications to geosciences. In: Nachtegael M, Van der Weken D, Kerre EE (eds), Soft computing in image processing: recent advances, Springer Verlag

48. Sinha AK (2006) Geoinformatics: data to knowledge. Geological Society of America Publ., Boulder, Colorado

49. Su T, Ozsoyoglu G (1991) Controlling FD and MVD inference in multilevel relational database systems. IEEE Transactions on Knowledge and Data Engineering 3:474–485

50. Sweeney L (1996) Replacing personally-identifying information in medical records, the scrub system. Journal of the American Medical Informatics Association 333–337

51. Sweeney L (1997) Weaving technology and policy together to maintain confidentiality. Journal of Law, Medicine and Ethics 25:98–110

52. Sweeney L (1997) Guaranteeing anonymity when sharing medical data, the datafly system. Journal of the American Medical Informatics Association 51–55

53. Sweeney L (1997) Computational disclosure control for medical microdata. In: Proceedings of the Record Linkage Workshop, Bureau of the Census, Washington, DC
54. Sweeney L (1998) Commentary: researchers need not rely on consent or not. New England Journal of Medicine 338(15)
55. Sweeney L (1998) Towards the optimal suppression of details when disclosing medical data, the use of sub-combination analysis. In: Proceedings of MEDINFO'98, International Medical Informatics Association, Seoul, Korea, North-Holland, p. 1157
56. Sweeney L (1998) Three computational systems for disclosing medical data in the year 1999. In: Proceedings of MEDINFO'98, International Medical Informatics Association, Seoul, Korea, North-Holland pp. 1124–1129
57. Sweeney L (1998) Datafly: a system for providing anonymity in medical data. In: Lin TY, Qian S (eds.) Database security XI: status and Prospects. Elsevier, Amsterdam
58. Torres R, Keller GR, Kreinovich V, Longpré L, Starks SA (2004) Eliminating duplicates under interval and fuzzy uncertainty: an asymptotically optimal algorithm and its geospatial applications. Reliable Computing 10(5):401–422
59. Trejo R, Kreinovich V (2001) Error estimations for indirect measurements: randomized vs. deterministic algorithms for 'black-box' programs. In: Rajasekaran S, Pardalos P, Reif J, and Rolim J (eds), Handbook on randomized computing. Kluwer, Dordrecht, 673–729
60. Vavasis SA (1991) Nonlinear optimization: complexity issues, Oxford University Press, New York
61. Willenborg L, De Waal T (1996) Statistical disclosure control in practice. Springer Verlag, New York
62. Wen Q, Gates AQ, Beck J, Kreinovich V, Keller JR (2001) Towards automatic detection of erroneous measurement results in a gravity database. In: Proceedings of the 2001 IEEE Systems, Man, and Cybernetics Conference, Tucson, Arizona, October 7–10, 2001, pp. 2170–2175
63. Xie H, Hicks N, Keller GR, Huang H, Kreinovich V (2003) An IDL/ENVI implementation of the FFT based algorithm for automatic image registration. Computers and Geosciences 29(8):1045–1055
64. Zelt CA, Barton PJ (1998) Three-dimensional seismic refraction tomography: A comparison of two methods applied to data from the Faeroe Basin. J. Geophysical Research 103(B4):7187–7210.

7

Fingerprint Recognition Using a Hierarchical Approach

Chengfeng Wang, Yuan Luo, Marina L. Gavrilova and Jon Rokne

University of Calgary, Calgary, Alberta, Canada
cwang,yluo,marina,rokne@cpsc.ucalgary.ca

This chapter introduces a topology-based approach to fingerprint recognition utilizing both global and local fingerprint features. It also describes a new hierarchical approach to fingerprint matching which can be used to accelerate the speed of fingerprint identification in large databases. The proposed matching scheme consists of two stages. In a coarse filtering stage, two fingerprint images are compared using the Singular Points method. Since the method does not resort to image enhancement and minutiae extraction, coarse matching is very efficient and can dramatically reduce the number of candidate fingerprints for minutia-level matching. During fine matching stage, Delaunay triangulation is used to speed up the minutia-level verification. This chapter also proposes to apply Radial Basis Functions to model fingerprint's elastic deformation, which greatly increases the system's tolerance to distorted images. Experimental results confirm that the proposed hierarchical matching algorithm achieves very good performance with respect to both speed and accuracy.

7.1 Introduction

Biometric technologies are automated methods of verifying or recognizing the identity of a person based on a psychological or behavioral characteristic. The recent developments in the areas of electronic banking, e-commerce, and security, has necessitated the development of reliable automatic personal authentication. Examples of physiological and behavioral characteristics commonly used for automatic identification include fingerprints, iris, retina, hand, face, handwriting, keystroke and voice. Fingerprint identification is one of the most reliable methods of personal authentication due to the number of properties, such as robustness, distinctiveness, availability and accessibility, that fingerprints posses. Compared to other biometrics, fingerprint recognition is a

C. Wang et al.: *Fingerprint Recognition Using a Hierarchical Approach*, Studies in Computational Intelligence (SCI) **57**, 175–199 (2007)
www.springerlink.com   © Springer-Verlag Berlin Heidelberg 2007

well-researched area, which experiences significant growth in the recent years. Existence of large fingerprint databases, for which it became very computationally expensive to repeat the matching procedures for each fingerprint image, drives the research further in this area. In this chapter, a scheme is presented which achieves much better matching performance for larger fingerprint databases in the presence of fingerprints with elastic distortions.

A fingerprint is formed from an impression of the pattern of ridges on a finger. A ridge is defined as a single curved segment, and a valley is the region between two adjacent ridges. In Fig.7.1 some of the ridges enter from the bottom-left of the image, loop around a common center point, and exit on the left. There are a lot of features in fingerprints, some of which are introduced below. Local ridge discontinuities, known as minutiae, have little effect on the

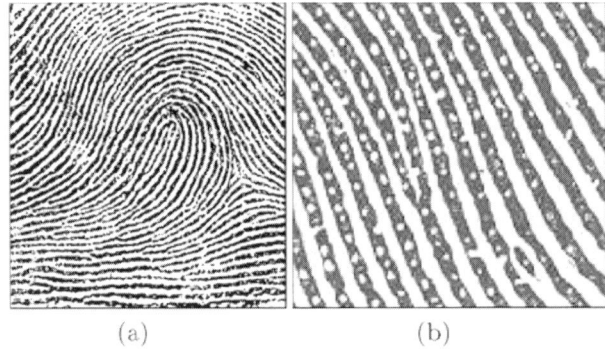

(a)   (b)

Fig. 7.1. (a) A fingerprint; (b) A magnified view of fingerprint ridge and valleys [2]

global ridge-valley pattern. However, it is the existence and location of these minutiae that embody much of a fingerprint's individuality. There are two kinds of minutiae: ridge endings and bifurcations. Ridge endings are places where ridges terminate and bifurcations are locations where a single ridge separates into two ridges (see Fig.7.2). Singular points have both global and local properties. There are two types of fingerprint singular points: cores and deltas. Locally, a core is the turning point of an inner-most ridge and a delta is a place where two ridges running side-by-side diverge. Core and delta points are illustrated by example (see Fig.7.2).

Automatic fingerprint identification systems typically include feature extraction and matching components. Here, we provide a brief overview about general methods of feature extraction, that will be more formally introduced in the later sections. To match two fingerprint images, one typically matched the feature pattern extracted from the images. Although many new features were proposed overtime for fingerprint identification, minutiae is still the most important one due to its good performance and distinctiveness. However, in practice, a fingerprint image may not always be well defined due to the ele-

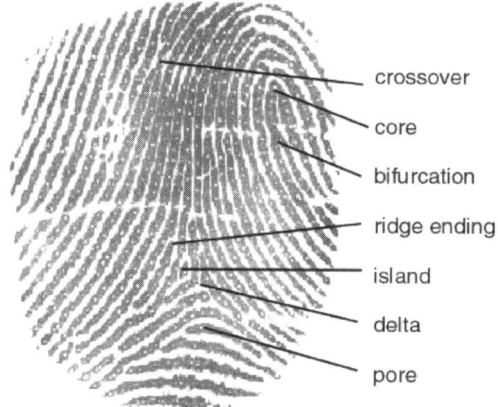

Fig. 7.2. Fingerprint features [2]

ments of noise that corrupt the clarity of the ridge structures. This corruption may occur due to variations in skin and impression conditions such as scars, humidity, dirt, and non-uniform contact with the fingerprint capturing device. Thus, image enhancement techniques are often employed to reduce the noise and to recover the topological structure of ridges and valleys from the noisy images. Fig.7.3 shows a complete flowchart for fingerprint image preprocessing and feature extraction systems. The description of the detailed steps for pre-processing and extrcation together with our contributions to the methodology will be presented in the Section 2 and Section 3 of this Chapter.

Majority of the fingerprint identification methods that appeared in the literature in recent years mainly differ in the features extracted from the fingerprint image. These features determine the performance of fingerprint matching that we coarsely classify into three families:

1. Correlation-based matching. Two fingerprint images are superimposed and the correlation between corresponding pixels is computed for different alignments [1];
2. Minutiae-based matching. This is the most popular and widely used technique. Minutiae are extracted from two fingerprints and stored as sets of points in the two dimensional plane [3, 15, 16];
3. Ridge-feature-based matching. These features can be the orientation map, the ridge frequency and so on [4].

Using the same amount of information, minutiae-based matching is considered to be the best approach, with the highest matching capability. Usually, algorithms for this family of matching stores minutiae's characteristics, including their coordinates, directions, shape and so on. In the method presented in this chapter, we employ minutiae features for local matching, and in addition propose to use some global features such as singular points and orientation

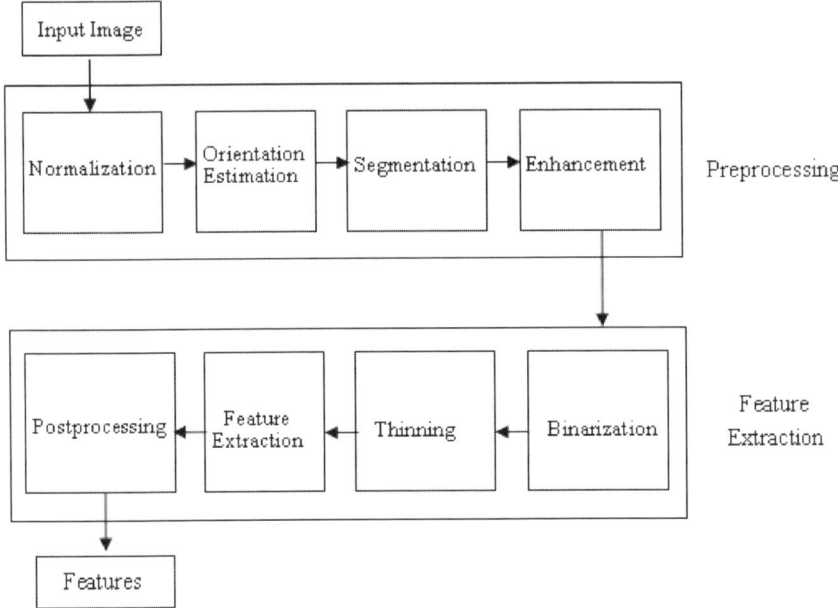

Fig. 7.3. The flowchart of our fingerprint feature extraction process

fields to improve the performance of current minutiae matching methods. The approach is radically different from other matching methods [3, 4, 15, 16], in that we apply the singular points and minutiae algorithms on two matching levels. This allows us to address the following challenges: (1) Missing and spurious features that can occur in poor quality images can be dealt with efficiently by our method; (2) Computationally expensive general minutiae-based matching methods are replaced by the fast and efficient two-level scheme which at the same time preserves matching accuracy; (3) The main difficulty for minutiae-base matching is the nonlinear deformation of a fingerprint which is explicitly modeled by our method in order to achieve a perfect alignment of the point sets.

Our Automatic Fingerprint Identification System can operate either in verification mode to comfirm the user who he says he is (one-to-one comparison) or identification mode to establish the identity of the user (one-to-many comparisons). The overall approach is described below. First, the system establishes whether the two fingerprints could be from the same finger, using the Singular Points (SPs) information extracted. We call this procedure the *Coarse Matching Stage*. As a global feature of fingerprints, SPs can be extracted much easier and quicker than minutiae points, which is why the SPs structure is used at the first filtering stage. By coarsely matching a fingerprint to the images stored in the database, approximately 30% of candidate finger-prints are excluded before the minutia-level verification stage, significantly improving the system performance. If a fingerpint and a canditate fingerprint from the database pass the first filtering stage, a topology-matching algorithm is applied at the *Fine Matching Stage*, where the two fingerprints are matched based on their local minutiae features. The method is based on building the Delaunay triangulation for the set of minutiae points, the fundamental com-putational geometry data structure that provides efficient representation for nearest-neighbour information. Compared with the traditional minutia match-ing methods, the proposed topology matching method differs in three main aspects. First, we use the Delaunay triangulation edges rather than minutiae or triangles build on minutiae points as a matching index to improve accuracy. Second, we apply a deformation model, which helps to deal with elastic fin-ger deformations that present significant challenges in fingerprint verification and thus improve reliability. Third, we apply a maximum bipartite matching scheme to achieve a better matching performance and improve the accuracy.

In the following sections, we describe the developed hierarchical matching algorithm in detail. Section 2 presents the main steps of *Coarse Matching Stage*, including Singular Points extraction and matching method based on the structure of SPs. The novel topology-matching algorithm is described in Section 3 as part of the *Fine Matching Stage*. Experimental results are pro-vided in Section 4. Finally, Section 5 contains the summary and conclusions.

7.2 Coarse Fingerprint Matching

The purpose of coarse fingerprint matching is to reduce the number of matches for fingerprint identification at the minutia-level. The traditional way to achieve this is fingerprint classification [5,7]. Basically, fingerprints are divided into several principal classes, such as: Arch, Tented Arch, Left Loop, Right Loop and Whorl. Then, the identification procedures are applied to one par-ticular class, which obviously reduces the number of candidate fingerprints. Unfortunately, the problem with this classification technique is that the num-ber of principal classes is small and the fingerprints are not uniformly distrib-uted (31.7%, 33.8%, 27.9%, 3.8%, 2.9% for R, L, W, A and T) [8]. Moreover, there are many fingerprints whose exclusive membership can not be reliably determined even by human experts.

Based on the recent results on singular points [9,10], we developed a dif-ferent approach for coarse fingerprints matching . The main steps of coarse matching algorithm include: foreground segmentation, singular points extrac-tion and matching. Compared with the traditional classification method, the proposed coarse matching algorithm is more precise and efficient, as will be demonstrated in the experimental section.

7.2.1 Fingerprint Foreground Segmentation

An important step in an automatic fingerprint recognition system before SPs extraction is the segmentation of fingerprint images (see Fig.7.3). A captured fingerprint image usually consists of two components, called the *foreground* and the *background*. The foreground is the component that originates from the contact of a fingertip with the sensor. The noisy area at the borders of the image is called the background. Accurate segmentation is especially important for the reliable extraction of features. Most feature extraction algorithms would not perform well when applied to the noisy background [2]. Therefore, the main goal of the segmentation is to discard the background, and thus reduce the number of falsely detected features.

Typical segmentation methods require computing the orientation fileds and coherence of each pixel, which is an expensive procedure [4, 15]. To better serve the purpose of fast coarse matching, we introduce a segmentation method based on an improved Canny Edge detector [20] and morphological operators. The main steps are as follows:

1. Normalization: Normalize the original image I to image G with a desired mean M_0 and variance VAR_0, which are optimized in different fingerprint databases by experimental data. The actual mean and variance of the original image is M and VAR. The normalization can be described by:

$$G(i,j) = \begin{cases} M_0 + \sqrt{\frac{VAR_0 \times (I(i,j) - M)^2}{VAR}} \ , & if \ I(i,j) > M \\ M_0 - \sqrt{\frac{VAR_0 \times (I(i,j) - M)^2}{VAR}} \ , & otherwise \end{cases} \quad (7.1)$$

After normalization, we use Gaussian smoothing to reduce the influence of noise.

2. Noise Reduction: Compute the gradient altitude

$$M(i,j) = \sqrt{\partial_x^2(i,j) + \partial_y^2(i,j)} \quad (7.2)$$

and the gradient orientation at each pixel,

$$O(i,j) = tan^{-1} \frac{\partial_y(i,j)}{\partial_x(i,j)} \quad (7.3)$$

where the gradients $\partial_x(i,j)$ and $\partial_y(i,j)$ are the x and y components of the gradient.

3. Non-maxima suppression: we divide the gradient orientation $O(i,j)$ into four directions (labeled as Fig.7.4), and each direction has its own orientation template for comparison. For example, if $O(i,j)$ belongs to district 2, we compare its gradient altitude $M(i,j)$ with its two neighbors in district 2. If $M(i,j)$ is not the local maximum, assign $M(i,j)$ to zero; otherwise $I(i,j)$ is a potential edge pixel.

2	3	4
1	$I(i,j)$	1
4	3	2

Fig. 7.4. Orientation district template

4. Edge hysteresis: we obtain the histogram of those potential edge pixels in non-maximum suppressed gradient matrix $M(i,j)$. Due to its double-apex shape, Otsu thresholding approach [6] is utilized to determine the high threshold $h1$ and the low threshold $h2$ is set to be $0.5 \times h1$. Once the two thresholds are determined, we follow the original hysteresis process of Canny algorithm [20] to extract the initial foreground boundary of the fingerprint image.

5. Boundary refinement: morphological operators are applied as the post-processing steps to refine the initial foreground boundary. First, hole blocks that are incorrectly assigned to the foreground are removed by means of an *opening* operation. Next, noise blocks that are incorrectly assigned to the background are removed by a *closing* operation.

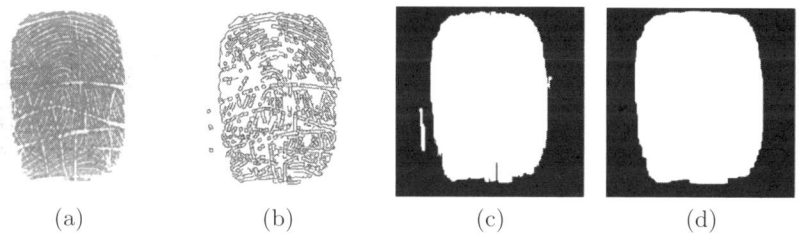

| (a) | (b) | (c) | (d) |

Fig. 7.5. Results of segmentation algorithm on a fingerprint (300×300). (a) Input fingerprint image; (b) Initial result of Canny Edge segmentation; (c) Foreground (white area) before morphology; (d) Foreground after morphology.

We record the edge pixel position as the foreground of fingerprint images. Fig.7.5 shows the results of our segmentation algorithm. Visual inspection demonstrates that this method provides satisfying results.

7.2.2 Singular Points Extraction

Once the foreground is segmented from the original fingerprint image, Singular Points are only extracted from the blocks that belong completely to the foreground areas. There are two types of Singular Points, core and delta, which denote the discontinuities in the orientation field in fingerprint images [11].

We locate every 3×3 block area containing SPs by calculating Poincare Index (PI) in block orientation field. PI represents the accumulation of orientation changes in a counter-clockwise close contour around an interested area (pixel or block). The value of PI will be π if there exists a core in this area, $-\pi$ if there is a delta, and 0 if no SPs exists [19]. In each block area, the type of the SPs is determined and if it is a core, the orientation of it will be calculated. Then, SPs are extracted according to local intensity gradient. The specific algorithm is presented below.

Computing Orientation Fields

We use the standard method presented in [5] to compute the orientation fields of fingerprints. We divide the input fingerprint image into 15×15 blocks.

$$\theta(i,j) = \frac{1}{2}\tan^{-1}\left(\frac{\sum_{u=i-\lfloor W/2\rfloor}^{i+\lfloor W/2\rfloor}\sum_{v=j-\lfloor W/2\rfloor}^{j+\lfloor W/2\rfloor} 2\partial_x(u,v)\partial_y(u,v)}{\sum_{u=i-\lfloor W/2\rfloor}^{i+\lfloor W/2\rfloor}\sum_{v=j-\lfloor W/2\rfloor}^{j+\lfloor W/2\rfloor}(\partial_x^2(u,v)-\partial_y^2(u,v))}\right) \quad (7.4)$$

where ∂_x, ∂_y are the same x and y component of gradient that we get during segmentation. Also, the Gaussian smoothing operation is performed at the block level (the filter size is set to 21×21 according to experimental data). (Fig.7.6).

Fig. 7.6. Fingerprint image (the left image) and block orientation field (the right image)

PI Calculating and SPs Position Locating

Next, the PI is calculated for those blocks in the foreground of the image. In each block, the PI is defined and computed as follows:

$$PI(i,j) = \frac{1}{2\pi}\sum_{k=0}^{N-1}\Delta(k) \quad (7.5)$$

$$\Delta(k) = \begin{cases} \delta(k) & \text{if } |\delta(k)| < \frac{\pi}{2} \\ \pi + \delta(k) & \text{if } \delta(k) < -\frac{\pi}{2} \\ \pi - \delta(k) & \text{otherwise} \end{cases}$$

$$\delta(k) = \theta(X(k'), Y(k')) - \theta(X(k), Y(k))$$
$$k' = (k+1) \bmod N$$

where $\theta(i,j)$ is the orientation of the block (i,j), and $X(k)$, $Y(k)$ are the coordinates of the blocks that are in the counter-clockwise of N blocks. In our experiment, N stands for its 8-neighbor blocks. Since different types of SPs (core and delta) have different orientation fields (Fig.7.8), we can determine whether SPs is existed in the block as well as the SP's type according to the block PI : if the PI is $1/2$ (Fig.7.7(a)), this block is a core block; if the PI is $-1/2$ (Fig.7.7(b)), this block is a delta block; if this block contains two core as a whorl, the PI is 1 (Fig.7.7(c)); if no SPs existed, the PI would be 0 (Fig.7.7(d)).

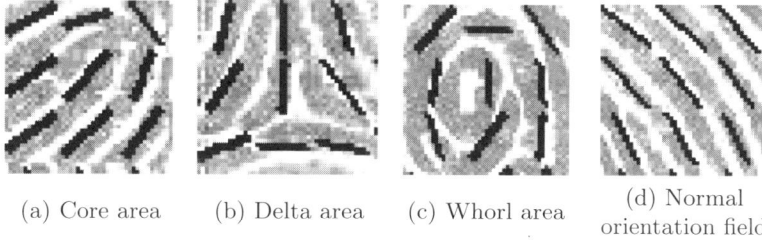

(a) Core area   (b) Delta area   (c) Whorl area   (d) Normal orientation field

Fig. 7.7. Block orientation field around different types of SPs.

If there exists a SPs in the block (i,j), we will locate it to a pixel-level precision within the block. Taking the similar approach proposed in [11], we treat the pixel whose directional gradients equal to zero as the exact location of SPs. However, in our system, the time-consuming pixel-level detection is only performed in a block which contains SPs, not on the whole fingerprint image as [11] did. And we do not need to take additional steps to determine the type of SPs in pixel-level, because SPs type is already determined at the block level. All these improvements lead to significant computational efficiency.

Singular Point Orientation Estimation

If the type of SPs is core, we propose to use a Block Orientation Template to coarsely estimate the orientation of the core. The detailed method is described below:

1. Select the most coherent block from the 8 neighbors of the considered block containing a core. This is achieved by computing the difference between the orientation of each neighbor block and the orientation in

the corresponding position of a Block Orientation Template (shown in Fig.7.8(b)). The block with the least difference is marked (shown in the Fig.7.9(a), 7.9(b) as gray block).

2. Calculate the orientation of the core according to orientation in both the marked block and adjacent blocks to the marked one. If the marked block is in the 4-neighborhood of the block containing a core, we use the mean value of the orientations of six blocks as the orientation of the core (see Fig.7.9(b); otherwise the mean value of the orientations of four blocks is used (see Fig.7.9(a)).

3. The orientation of the core is normalized to the range $(-\pi/2, \pi/2)$ as the last step.

We take into account both the orientation of the marked block and of its adjacent blocks. Since the orientations of ridge lines are more consistent than those in the block containing the core, a robust computation of the orientation of the core can be attained in this way.

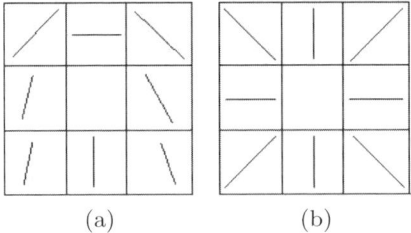

(a)   (b)

Fig. 7.8. (a) Typical block orientations in a 3×3 blocks with the center one containing a core, (b) Block orientation template

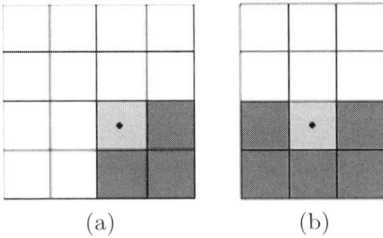

(a)   (b)

Fig. 7.9. Blocks involved when computing the orientation of core. The block containing a core is marked with a black dot, the marked block is filled with gray, and those adjacent blocks considered are filled with black. (a) the case if the marked block is in the 8-neighborhood but not in the 4-neighborhood of the block containing a core (b) the case if the marked block is a 4-neighbor of the block containing a core

7.2.3 Singular Points Matching

Based on the number, the position and the orientation of SPs, fingerprints are matched at SPs level. If these two fingerprints even fail to be matched at SPs level, we do not need to extract their minutia points for detail matching, which skips the most time-consuming step for fingerprint identification. Fig.7.10 shows the four basic structures of SPs, our similarity measurement is applied based on the distance between two SPs, the angle difference between the orientation of the core and the line connecting two SPs. More specific coarse matching rules are presented in Table 7.1:

SPs No.	2 Cores	1 Core & 1 Delta	1 Core
2 Cores	*SM	*DM	*DM
1 Cores & 1 Delta	*DM	*SM	*DM
1 Core	*DM	*DM	PASS

Table 7.1. Coarse Matching Rules

- *SM (Similarity Measure): if the differences of distance between two SPs and the angle between the orientation of core with connecting line are less than two specific thresholds, make a positive decision that these two fingerprints match.
- *DM (Distance Measure): if the minimum distance from the core to the background in one fingerprint image is greater than the distance between two cores in another fingerprint image, make a negative decision that these two fingerprints do not match.

The basic principle of our coarse matching is that we only reject those fingerprints that definitely are not of the same type. This lets all fingerprints with similarities pass. For example, in Fig.7.10, structure (a) and (c) might possibly have to be matched, because there might be a delta in (a), which is out of the fingerprint foreground due to the enrolment position. Here we should compare the minimum distance D_{min} from that core to the background with the distance D_{sp} between two SPs: if D_{min} is greater than D_{sp}, which means that the second core cannot exist in that structure, we reject these two fingerprints. If more than two cores are detected in a fingerprint, it is usually due to the influence of noise. All these fingerprint images including those without any SPs are passed to the minutiae-level verification.

7.3 Topology-based Fine Matching

The motivation for seeking a topological description of fingerprints is provided by the elasticity of the human skin since successive rolled impressions from the

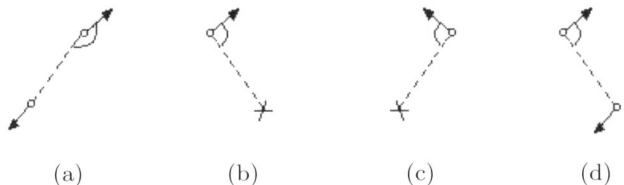

(a)   (b)   (c)   (d)

Fig. 7.10. Four structures of Singular Points (Core marked by circle, delta marked by cross)

same finger will invariably suffer a degree of relative distortion (translation, rotation and stretching). A topology based system would be less affected by the negative effects of plastic distortion.

Our developed hierarchical fingerprint matcher utilizes two distinct sets of fingerprint information: singular points and minutiae points. In above section, we proposed a fingerprint coarse matching scheme to avoid matching all pairs of fingerprints in a large database. Many potential candidate fingerprints can be excluded during the *Coarse Matching Stage*. However, for the fingerprints that are passed, the SPs are not distinctive enough to identify which one is from the same finger. In the fingerprint fine matching stage, we proposed a novel topology-based method for identifying such fingerprints.

In order to determine if two fingerprint images are from the same finger, we have to compare them based on their minutiae. We align the input fingerprint with the template fingerprint represented by its minutia pattern. But even if an input image comes from the same finger as the template image in the database, there are transformations such as translation, rotation and scaling that might make the matching difficult. To match two fingerprint images, we need to estimate these transformation parameters to align input minutiae with the template minutiae (see Fig.7.11).

Under a rigid transformation $(\Delta\theta, \Delta x, \Delta y)$, a point from the input image can be transformed to the corresponding point of the template image after rotating $\Delta\theta$ and translating $(\Delta x, \Delta y)$. The assumption is that the scaling factor between input and template images is identical since both images are captured with the same device.

The flow diagram for the fingerprint fine matching algorithm is shown in Fig.7.12. We first use edges of the Delaunay triangle as a matching index. Second, we use a deformation model, which helps to deal with the elastic finger deformations that sometimes are very damaging for correct fingerprint verification. Third, to achieve a better performance, we consider singular points and minutiae sets for matching and apply a new matching scheme. The key benefit for using topology-based approach for fingerprint matching is an exceptionally good performance of the technique even in the presence of bad quality input images. We will discuss this in the following sections.

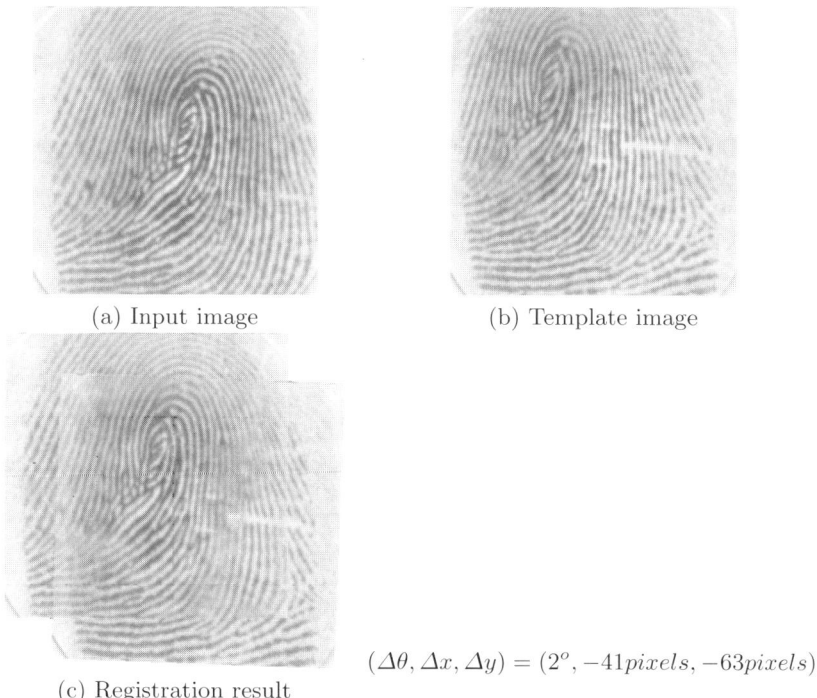

(a) Input image

(b) Template image

(c) Registration result

$(\Delta\theta, \Delta x, \Delta y) = (2^o, -41pixels, -63pixels)$

Fig. 7.11. Registration of fingerprint image

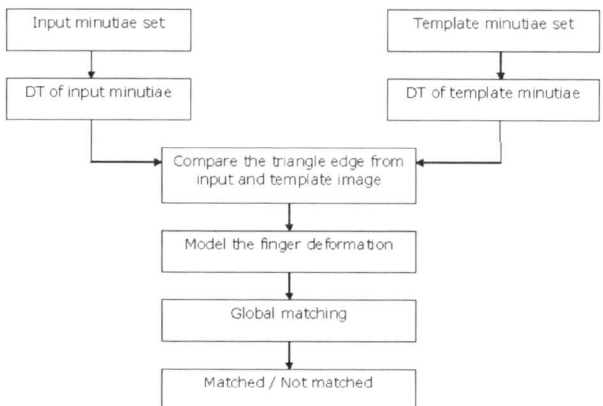

Fig. 7.12. Fingerprint fine matching flowchart

7.3.1 Delaunay Triangulation of Minutiae Set

The Voronoi Diagram and Delaunay Triangulation (DT) are closely related data structures [18], that can be used to describe the topology of the fingerprint, which is considered to be the most stable information for fingerprint matching purposes (Fig. 7.13).

Voronoi region associated with a feature is a set of points closer to that feature than to any other feature. Given a set S of points $p_1, p_2, ..., p_n$, the Voronoi diagram decomposes the 2D space into regions around each point p_i such that all points in the region around p_i are closer to it than they are to any other point from S. Let $V(S)$ be a Voronoi diagram of a planar point set S. Consider the straight-line dual $D(S)$ of $V(S)$, i.e., the graph embedded in the plane obtained by adding a straight-line segment between each pair of points in S whose Voronoi regions share an edge. The dual of an edge in $V(S)$ is an edge in $D(S)$. $D(S)$ is a triangulation of the original point set, and is called the Delaunay triangulation after Delaunay who proved this result in 1934.

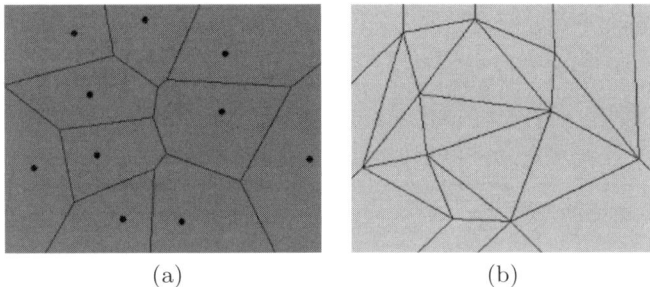

(a)   (b)

Fig. 7.13. The Voronoi Diagram and Delaunay Triangulation of a set of points

The purpose of fingerprint local matching is to determine the transformation parameters $(\Delta\theta, \Delta x, \Delta y)$ between an input fingerprint and a template fingerprint. Features such as minutiae, singular points, ridges and orientation field can be used to determine the transformation.

We attempt to acquire the best registration of two fingerprint images, by first assuming that they are from the same finger. To register two images, we propose to use the Delaunay minutiae triangle as an important topological structure of the fingerprint images. A novel method basinged on the Delaunay triangulation is applied to match minutiae pairs.

We proposed to use DT for the alignment of the minutiae set for several important reasons: (1) The DT is uniquely identified by the set of minutiae points. (2) Inserting a new minutiae point or dropping a minutiae point in a set of points in most cases affects the triangulations locally, which means the algorithm can tolerate some error in minutiae extraction. (3) The number of triangles in DT is only $O(n)$. (4) As a very useful topology structure, DT can tolerate some fingerprint deformation. Fig.7.14 shows a Delaunay minutiae triangulation of two corresponding images and it also shows the reason for using triangle edges to compare rather than the whole triangle. It is more likely to find a matched triangle edge pair than a whole triangle pair in corresponding images.

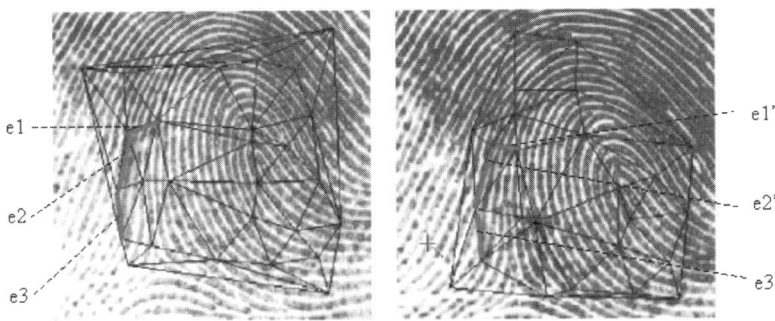

Fig. 7.14. Delaunay minutiae triangle (three triangle edges are matched, where $e1$, $e2$, and $e3$ match $e1'$, $e2'$, and $e3'$ respectively, but not a single whole triangle is matched)

Let $Q = ((x_1^Q, y_1^Q, \theta_1^Q)...(x_n^Q, y_n^Q, \theta_n^Q))$ denote the set of n minutiae points in the input image $((x,y)$:minutiae location; θ: orientation field of minutiae) and $P = ((x_1^P, y_1^P, \theta_1^P)...(x_m^P, y_m^P, \theta_m^P))$ denote the set of m minutiae points in template image. When two edges match successfully, their lengths and their angles related to the orientation field should be identical with some tolerance. The transformation $(\Delta\theta, \Delta x, \Delta y)$ is obtained by comparing the minutiae pairs of these two edges:

$$\Delta\theta = \theta_i^Q - \theta_j^P$$
$$\Delta x = x_i^Q - x_j^P \cos\Delta\theta + y_j^P \sin\Delta\theta \qquad (7.6)$$
$$\Delta y = y_i^Q - x_j^P \sin\Delta\theta - y_j^P \cos\Delta\theta$$

(In the two edges, minutia $m_i^Q = (x_i^Q, y_i^Q, \theta_i^Q)$ match minutia $m_j^P = (x_j^P, y_j^P, \theta_j^P)$.)

If one edge in the input image matches another edge in template image, we obtain one transformation satisfying the above criteria. Yet, we might have more than one matched triangle edge pairs. For example, there are three matched pairs in Fig.7.14.

For a certain range of translation and rotation dispersion, we put each transformation $(\Delta\theta_i, \Delta x_i, \Delta y_i)$ as a point in the transformation space, detect the peak where there are maximum number of transformations in the range, and record those transformations that are neighbors of the peak in transformation space.

Through comparing the minutiae Delaunay triangle edges, we can record a group of transformations $(\Delta\theta_i, \Delta x_i, \Delta y_i)$. These transformations should be identical, if the input image and the template image are from the same finger and there is no non-linear deformation. Yet, non-linear deformation always exists and can affect the fingerprint matching accuracy. We propose a new method to address the deformation problem in fingerprint and it can improve the matching accuracy in the following section.

7.3.2 Modeling Fingerprint Deformation

Fingerprint deformation problems arise because of the flexibility of the finger. Some elastic distortion will necessarily result from the skin of the finger not being a planar surface. Other distortions arise from the pliability of the underlying tissue. Consequently, pressing or rolling it against a flat surface induces distortions which vary from one impression to another. Such distortions lead to relative translations of features when comparing one fingerprint image with another.

A good fingerprint identification system will always compensate for these deformations. We develop a simple framework aimed at approximately quantifying and modeling the local, regional and global deformation of the fingerprint. We propose to use a Radial Basis Function (RBF), which represent a practical solution to the problem of modeling of a deformable behavior. The application of RBFs has been explored in medical image matching and image morphing [14, 17]. To the best of our knowledge, it has not been applied for fingerprint verification. In the followings, we propose a RBF based method and apply it to fingerprint matching.

When registering images from rigid regions, a spatial transformation is normally used in the form of a rigid or affine transformation. These transformations account for global differences, but ignore non-rigid deformations. When matching non-rigid images, it is important to use a spatial model which adequately represents the spatial differences that exist between them. If the selected transformation does not truly model the spatial differences, an overall accurate match will not be obtained, even when accurate control point correspondences are found by matching the DT edges in the local matching.

When two images are registered, we select a couple of corresponding minutiae point pairs and obtain a group of transformations from these corresponding minutiae point pairs. We call these points *control points*. For our fingerprint matching algorithm, the deformation problem can be described as: knowing the consistent transformations of some points which we call *control points* in the minutiae set of input image, how to interpolate the transformation of other minutiae which are not control points? We do not consider all the transformations obtained by the local matching. We pick those consistent transformations which form large clusters. This allows us to consider both global deformation and local deformation.

In the following, we briefly describe the proposed RBF based method in two dimensions. Given the coordinates of a set of corresponding points (control points) in two images: $\{(x_i, y_i), (u_i, v_i) : i = 1, ..., n\}$, determine function $f(x, y)$ with components $f_x(x, y)$ and $f_y(x, y)$ such that:

$$
\begin{aligned}
u_i &= f_x(x_i, y_i) \\
v_i &= f_y(x_i, y_i) \quad i = 1, ..., n
\end{aligned}
\tag{7.7}
$$

A spatial transformation is considered rigid if the distance between any two points is preserved. Rigid transformation can be decomposed into a translation and a rotation. In 2D, a simple affine transformation can represent rigid transformation as:

$$
f_k(\mathbf{x}) = a_{1k} + a_{2k}x + a_{3k}y \quad k = 1, 2
\tag{7.8}
$$

where $\mathbf{x} = (x, y)$. There are only three unknown coefficients in a rigid transformation. The RBF is used for a non-rigid transformation appropriate for fingerprint matching. In the two dimensional case, this transformation is determined by $n + 3$ coefficients in each dimension:

$$
f_k(\mathbf{x}) = a_{1k} + a_{2k}x + +a_{3k}y + \sum_{i=1}^{n} A_{ik}g(r_i) \quad k = 1, 2
\tag{7.9}
$$

The first three terms is an affine transformation. The last term is the sum of a weighted elastic or nonlinear basis function $g(r_i)$, which is related to the distance between $\mathbf{x}$ and the ith control point. The coefficients of the function $f_k(\mathbf{x})$ are determined by requiring that $f_k(\mathbf{x})$ satisfy the interpolation conditions:

$$
\begin{aligned}
f_1(\mathbf{x_i}) &= u_i \qquad \text{and} \\
f_2(\mathbf{x_i}) &= v_i \quad \text{for} \quad i = 1, 2, ..., n
\end{aligned}
\tag{7.10}
$$

where n is the number of control points. Giving n linear equations together with the additional compatibility constrains:

$$
\sum_{i=1}^{n} A_{ik} = \sum_{i=1}^{n} A_{ik}x_i = \sum_{i=1}^{n} A_{ik}y_i = 0 \quad k = 1, 2
\tag{7.11}
$$

These conditions guarantee that the RBF is affinely reducible, i.e. the RBF is purely affine whenever possible. There are $2(n+3)$ unknown coefficients and $2(n+3)$ constraints in Eq.7.9 and Eq.7.10. The $2(n+3)$ unknown coefficients of the basis function and the polynomial are solved in [14, 17].

The application of the deformation model can help us to match two corresponding images with a higher accuracy, as shown in Fig.7.15. Fig.7.15(c) shows the alignment of two images Fig.7.15(a) and (b) when we consider the transformation of input image to be a rigid transformation. We tried every transformation in the rigid transformation space and found the maximum

number of matching minutiae pairs between input image and template image
to be 6 (labeled by dash circle in Fig.7.15(c)). Circles denote minutiae of
input image after transformation, while squares denote minutiae of template
image. Knowing the transformation of five minutiae (*control points*) in the
input image, we apply the RBF to model the non-rigid deformation, which
is visualized by the deformed grid in Fig.7.15(d). The number of matching
minutiae pairs is 10 which greatly increases the matching scores of these two
corresponding images. Experiments presented in Fig.7.15 show that we can
detect more matching minutiae pairs if we regard there is a non-rigid transformation between input and template images, and apply the RBF to model
the deformation. A considerably better matching score can be obtained after
we apply this technique.

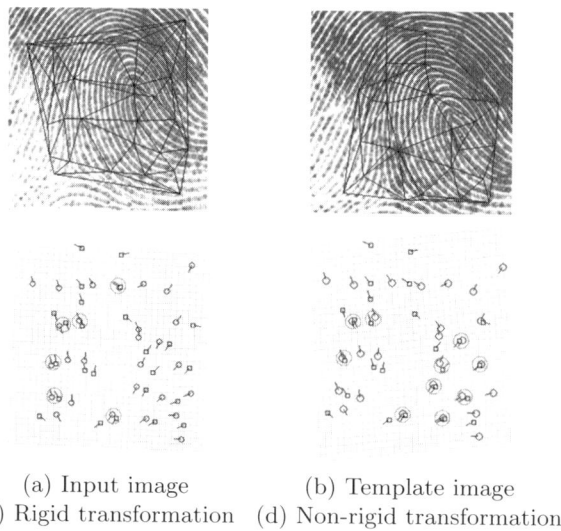

(a) Input image   (b) Template image
(c) Rigid transformation (d) Non-rigid transformation

Fig. 7.15. Comparison of rigid transformation and non-rigid transformation in
fingerprint matching

7.3.3 Maximum Bipartite Matching

In the above two sections, we used DT to register two fingerprint images,
and applied RBF to model the non-linear deformation of fingerprints. Fig.7.11
shows the registration result ignoring the non-linear deformation, and Fig.7.15
(d) show the registration result considering the non-linear deformation. Now,
we need to compare the fingerprints similarity after registration.

In this section, we discuss the fingerprint global matching. The purpose of
global matching is to find the number of matching minutiae of two fingerprint
images after registration. If the number of matched minutiae pairs is greater

than a threshold, we can conclude that these two images are corresponding images from the same finger. Based on theory about maximum flow problem, we propose a novel maximum bipartite matching scheme for this purpose. Maximum flow is a traditional problem to compute the greatest rate at which substance can be transferred from the source to the sink without violating any capacity constrains [13]. To the best of our knowledge, this is the first application of maximum flow to fingerprint minutiae matching.

We need to match minutiae pairs under a known transformation. If one minutia from the input image and one minutia from template image fall into the same tolerance box after transformation, they are defined as matched. However, to obtain an optimal pairing, which maximize the number of matched point pairs, is not as easy as it looks.

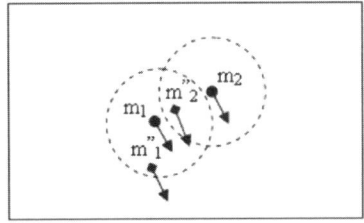

Fig. 7.16. Strategy of matching minutiae pairs. In this example, if m_1 was matched with $m_2^{''}$ (the closest minutiae), m_2 would remain unmatched; however, pairing m_1 with $m_1^{''}$, allows m_2 to be matched with $m_2^{''}$, thus maximizing the matching pairings.

The most important rule of matching feature points is to guarantee that one minutia from the input image can match to at most one minutia from the template image. To comply with this constraint, one can mark the minutiae that have already been matched to avoid matching them twice or more. However, it is hard to find the optimal pairing of the feature points. For example, in Fig.7.16. the best pairing is the configuration that can maximize the final number of matched minutia pairs. A more sophisticated method should be used to obtain this optimum pairing.

We propose to utilize a bipartite weight graph $B = (Q, P)$ (see Fig.7.17) to match the minutiae pairs, where Q denotes the set of n minutiae in the input image and P denote the set of m minutiae in template image. For minutiae q in Q and minutiae p in P, if their locations and angles are identical with some tolerance, an edge is built between them. The capacity for these edges is set 1. For edges from source s to q and p to sink t, we set the capacity to 1. In this setting, the problem of finding the maximum number of matched point pairs between the input and template images is turned into finding the maximum flow in this bipartite graph. We can use Ford-Fulkerson algorithm [13] to find the maximum flow in such bipartite graph.

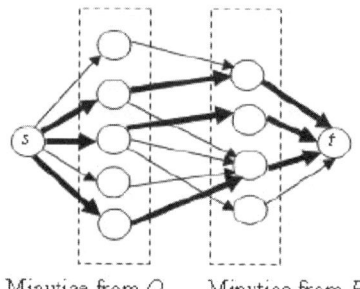

Minutiae from Q Minutiae from P

Fig. 7.17. Point pattern matching problem. The edge from s to Q has capacity 1. The edge from P to t has capacity 1. Minutiae from Q and P are connected if these two minutiae are paired, and capacity of that edge is 1. The occurrence arrow corresponds to those in a maximum flow or maximum matching pairs of the input and template images.

7.4 Experimental Results

The system was implemented using C++ programming environment and tested on a Pentium 4 2.8GHZ CPU, 512 RAM computer. Experiments have been done on three different fingerprint databases. Database1 is from FVC2000 for SPs detection. Database2 includes 448 fingerprint images of size 300×300 captured by a Veridicom COMS sensor. Database3 is from Biometric System Laboratory at University of Bologna. Database3 consists of 21×8 fingerprint images, where there are 21 fingers and each fingers contains 8 images captured in different time.

In order to compare our SPs detection results with other algorithms, tests are first performed in the second database of FVC2000. Fig.7.18 shows some samples of SPs detection. Comparison is made among the performances of algorithm in [11] and the developed approach, see Table 7.2. Obviously, our SPs detection methods are much better than the methods in [11] both in efficiency and accuracy

Alg.	No. of False SPs	Ratio of False SPs	Ratio of Missed SPs	Proc Time
[11]	2.67	0.73	0.55	1.66 s
Ours	0.31	0.11	0.17	0.45 s

Table 7.2. Performance comparison between [11] and the proposed algorithms on database1

Coarse matching is tested on database2. Our test strategy is: True Matching algorithm is applied to fingerprint images from a same finger to measure the FRR; False Matching algorithm is applied to the fingerprints from different fingers to measure the FAR. We conduct the True Matching 1016 times,

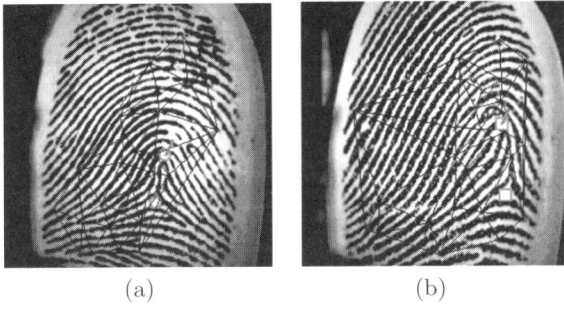

(a)   (b)

Fig. 7.18. Results of SPs detection (the orientation of the core is indicated by a line) and Delaunay minutiae triangulation

with 36 regarded as different fingerprints; False Matching 2016 times, 629 times to successfully reject fingerprints from different fingers, that is to say, to search a fingerprint in a database, 31.2% fingerprints can be excluded in the coarse matching level.

When identifying a fingerprint in a large database, we can save time by excluding part of them at the coarse matching level. In the *Fine Matching Stage*, we match the input image with all other fingerprints in the database on minutiae-level except the excluded part. Table 7.3 shows the computational

Delaunay Triangulation	Deformation	Global Matching	Total
$O(nlogn)$	$O(n^2)$	$O(n^2)$	$O(n^2)$

Table 7.3. Computation complexity of our *Fine Matching Stage*

complexity of fingerprint matching on minutiae level. There are three steps in *Fine Matching Stage*: local matching, deformation model and global matching. Let n refer to the number of minutiae. The computational complexity for Delaunay triangulation construction by the sweep-line method is $O(nlogn)$. In the worst case (there are n control points), the calculation of RBF function and registration takes $O(n^2)$ time. Finally, it takes $O(n^2)$ time to match point pairs in global matching. We sum the computation complexity of them. Thus, the computation complexity of our algorithm is $O(n^2)$. It is $O(n^3)$ for most minutiae or minutiae and ridge based methods [15, 16]. Experimental results also prove that our algorithm performs faster than standard methods. Average time of one matching is about 27ms (feature extraction time is not included). Average time cost is 98ms for standard methods [15, 16] using the same computer and the same minutiae sets.

When evaluating a biometric identification system, there are two important criteria. The FAR (False Accepted Rate) is a statistical measurement of the number of impostors likely to be accepted by a biometric system. The FRR (False Refused Rate) is a statistical and empirical measurement of the likelihood of genuine users being rejected by a system. We tested our hierarchical matching algorithm on database3 from University of Bologna. The FRR and FAR are 6.53% and 0.16% respectively for our hierarchical approach. We also test the other minutiae based fingerprint identification algorithms [15,16].

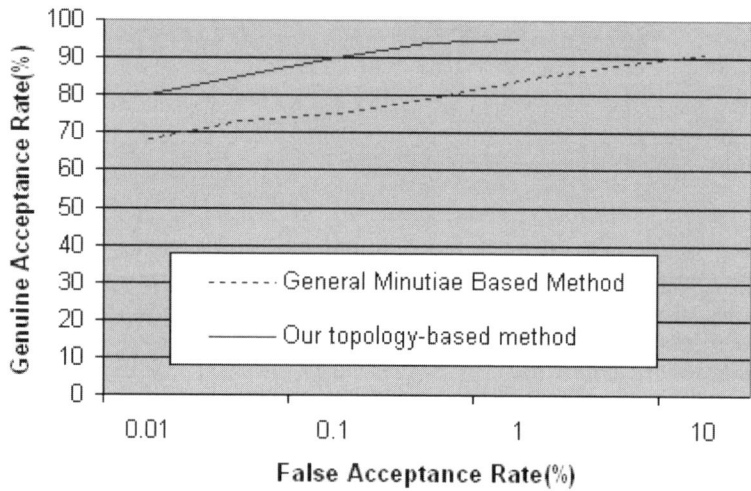

Fig. 7.19. ROC curves of general method vs. our method

Receivers operating characteristic (ROC) curves can provide the relationship between FAR and FRR. For convenience of comparison, we use genuine acceptance rate, which is an often used criterion for testing of matching algorithms. It is equal to $1 - FRR$. The comparison between our algorithm and the general minutiae based matching method are shown in Fig.7.19. Dotted curve and solid curve represent general minutiae-based method and our topology-based method respectively. To explain the figure, we can take some points as an example. When FAR is 0.1% for both methods, the genuine acceptance rate of our method is 88%, and that of general minutiae-based method is 74%. The performance of our matching algorithm is much better.

The following tests are directed to study our methods tolerance to distortion. One of the main difficulties in matching two fingerprint samples of the same finger is to deal with the non linear distortions (see Fig.7.20), often produced by an incorrect finger placement over the sensing element. This would make a global rigid comparison not feasible. To illustrate that our algorithm is robust against the non-linear distortion, we conduct further experiments on the distorted images. Thus, we apply a mathematical model to model the real distortion [12].

We assumed that the original images in test database were images without non-linear distortion. In the plastic distortion model, there are three distinct

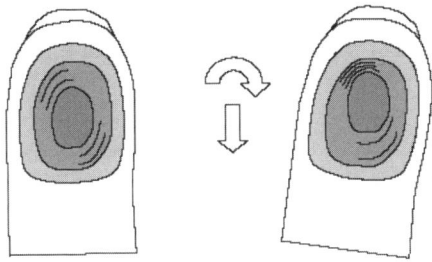

Fig. 7.20. View of a finger before and after the application of traction and torsion forces [12]

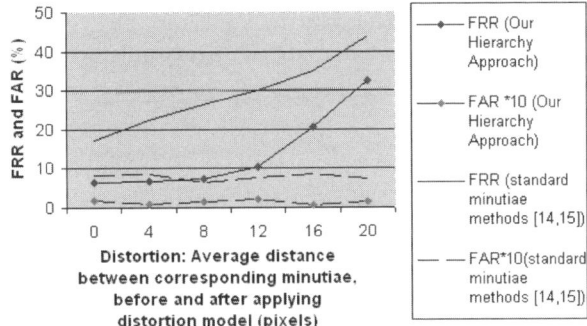

Fig. 7.21. Performance of topology methods vs. standard methods [15, 16] under distortion

regions in the fingerprint (see the three gray areas in Fig.7.20). A close-contact region does not allow any skin slippage. For minutiae points in other two regions, their locations will be changed. The average distance in Fig.7.21 represents the average distance change of minutiae points in pixels after we distort the images. When there is no distortion, the location of minutiae points does not change and the average distance is 0. Fig.7.21 shows that when the distortion is not very significant, i.e., average distance is less than 8 pixels, the accuracy of our matching algorithm remains almost the same. However, the FRR of the standard minutiae methods [15, 16] will increase in a much faster manner than ours. In other words, our algorithm is more robust against the non-linear distortion than the standard methods.

7.5 Conclusions

This chapter introduced a new hierarchical approach for fingerprint identification by using both the global and local features of fingerprints. During *Coarse Matching Stage*, SPs information is extracted as the global feature of that fingerprint. Without image enhancement and minutiae extraction, coarse matching is computationally efficient and can greatly reduce the number of candidate fingerprints for minutia-level matching. During *Fine Matching Stage*, we use the local features: minutiae. They are determined using a Delaunay triangulation based technique which leads to triangle matching a fast process. This overcomes the non-linear deformation problems. To overcome the relative deformation that is present in the fingerprint image pairs, we also propose a novel RBF model that is able to deal with elastic distortions. The proposed maximum bipartite matching schema can optimize the global matching. Our experiments show that fingerprint images can be well matched using the hierarchical matching method even on fingerprint database with some nonlinear distortion. The investigation of other features and their application in coarse matching to exclude more candidate fingerprints in large database will speed up the fingerprint recognition. The topological features, methods and their influence on accuracy of fingerprint recognition can be the direction for future research.

References

1. Asker M.B, Gerben T.B., Verwaaijen, Sabih H.G., Leo P.J. and Berend J.Z. 2000, A Correlation-Based Fingerprint Verification System, Proceedings of ProRISC 2000 Workshop on Circuits, Systems, and Signal Processing, the Netherlands, pp. 205-213.
2. Maltoni, D. Jain A.K., and Maio D. 2003, Handbook of Fingerprint Recognition, Springer-Verlag.
3. Bebis G., Deaconu T., and Georiopoulous M. 1999, Fingerprint identification using Delaunay triangulation, ICIIS99, Maryland, pp. 452-459.
4. Jain A.K., Salil P., Hong L., and Pankanti S. 2000, Filterbank-Based Fingerprint Matching, IEEE Transactions On Image Processing, vol. 9, pp. 846-859.
5. Kamijo M. 1993, Classifying Fingerprint Images using Neural Network: Deriving the Classification State, IEEE International Conference on Neural Network, vol. 3, pp. 1932-1937.
6. Otsu N. 1979, A Threshold Selection Method from Gray-Level Histograms, IEEE Transactions on Systems, Man, and Cybernetics, vol. 9, pp. 62-66.
7. Rao K. and Balck K. 1980, Type Classification of Fingerprints: A Syntactic Approach, TPAMI, vol. 2, pp. 223-231.
8. Bhanu B. and Tan X. 2003, Fingerprint Indexing Based on Novel Features of Minutiae Triplets, TPAMI, vol. 25, pp. 616-622.
9. Wang C.F. and Gavrilova M.L. 2004, A Multi-Resolution Approach to Singular Point Detection in Fingerprint Images, Proceedings of the International Conference on Artificial Intelligence, vol. 1, Las Vegas, Nevada, USA, pp. 506-514.
10. Wang F., Zou X., Luo Y., and Hu J. 2004, A Hierarchy Approach for Singular Point Detection in Fingerprint Images, Proceedings of the first International Conference on Biometric Authentication (ICBA04), vol. 1, HongKong, pp. 359-365.
11. Ramo P., Tico M., Onnia V., and Saarinen J. 2001, Optimized singular point detection algorithm for fingerprint images, Proceedings. International Conference on Image Processing, vol. 3, pp. 242-245.

12. Cappelli R., Maio D. and Maltoni D. 2001, Modelling Plastic Distortion in Fingerprint Images, ICAPR, LNCS 2013, pp. 369-376.
13. Cormen T., Leiserson C., Rivest R., and Stein C. 2002, Introduction to Algorithm, The MIT Press.
14. Fornefett M., Rohr K. and Stiehl H.S. 2001, Radial basis functions with compact support for elastic registration of medical images, Image and Vision Computing, vol. 19, pp. 87-96.
15. Jain A.K., Hong L., and Bolle R. 1997, On-line fingerprint verification, TPAMI, vol. 19, pp. 302-314.
16. Ratha N.K., Karu K., Chen S., and Jain A.K. 1996, A Real-Time Matching System for Large Fingerprint Databases, TPAMI, vol. 18, pp. 799-813.
17. Wang C.F. and Hu Z.Y. 2003, Imange Based Rendering under Varying Illumination, the Journal of High Technology Letters, vol. 9, pp. 6-11.
18. Okabe A., Boots B., and Sugihara K. 1992, Concepts and Applications of Voronoi Diagrams, Wiley Publishing, Chichester, England.
19. Karu K. and Jain A.K. 1996, Fingerprint Classification, Pattern Recognition, vol. 18, pp. 389-404.
20. Canny J. 1986, A Computational Approach to Edge Detection, TPAMI, vol. 8, pp. 679-698.

8

Smart Card Security

Kostas Markantonakis[1], Keith Mayes[1], Michael Tunstall[1], and Damien Sauveron[2] Fred Piper[3]

[1] Smart Card Centre, Information Security Group, Royal Holloway, University of London, Egham, Surrey TW20 0EX, UK
   (k.markantonakis, keith.mayes, m.j.tunstall)@rhul.ac.uk
[2] XLIM — UMR CNRS 6172, University of Limoges, 123 avenue Albert Thomas — 87060 LIMOGES Cedex, FRANCE
   damien.sauveron@unilim.fr
[3] Information Security Group, Royal Holloway, University of London, Egham, Surrey TW20 0EX, UK
   f.piper@rhul.ac.uk

In recent years smart cards have become one of the most common secure computing devices. Their uses include such diverse applications such as: providing secure wireless communication framework, banking and identification. Direct threats to smart card security can be invasive (attacks that alter the chip inside the card), analysis of a side channel, induced faults, as well as more traditional forms of attack. This chapter will describe the various attacks that can be applied to smart cards, and the subsequent countermeasures required in software to achieve a secure solution. Invasive attacks are considered beyond the scope of this chapter and can be mitigated with software countermeasures. A case study on the various generations of the European mobile telephone networks is given as an example of how the deployment of countermeasures has changed due to the various attacks described in this chapter.

8.1 Introduction

Smart cards are used to ensure secure execution and secure storage. A secure smart card has to guarantee that secret information cannot be retrieved or modified by an unauthorised entity. This information could be secret keys used in cryptographic algorithms, an e-purse balance, etc. Smart cards are generally not vulnerable to classical software attacks implemented against PCs, but their unique form as led to different methods of attack. These attacks are briefly described below; a more thorough description including a description of the necessary countermeasures is given in subsequent sections.

K. Markantonakis et al.: *Smart Card Security*, Studies in Computational Intelligence (SCI) **57**, 201–233 (2007)
www.springerlink.com   © Springer-Verlag Berlin Heidelberg 2007

Side Channel Attacks: Sensitive systems that are based on smart cards use
well-known cryptosystems. Generally, these cryptosystems have been subject to rigorous mathematical analysis in the effort to uncover weaknesses
in the system. The cryptosystems used in smart cards are therefore not
usually vulnerable to these types of attacks. Since smart cards are small,
physical objects that can actually be carried around in ones pocket, adver-
saries turned to different mechanisms of attack. Side channel analysis is a
class of attacks that seek to deduce information in a less direct manner.
This is achieved by extracting secret information held inside devices, such
as smart cards, via the monitoring of information that leaks naturally
during its operation.
The first attack of this type was a timing attack against RSA and
some other public-key cryptographic primitives [42]. This was followed
by attacks using the power consumption [43] and electromagnetic emana-
tions [28] as a side channel to extract information.
Fault Attacks: Since the publication of an fault attack against the RSA in
1997 [12] fault analysis has come to the foreground as a possible means of
attacking devices such as smart cards. The vast majority of papers written
on this subject involved very specific faults (usually a bit-flip in a specific
variable) that are extremely difficult to produce. However, there have been
several attacks published that have enough freedom in the type of fault
required that they can be realised with current fault injection methods.
Fault attacks have been implemented against chips that could be used in
smart cards [4]. This has meant that countermeasures against this type
of attack need to be implemented in embedded devices to protect against
this class of attack.
Multi-Application Security: Modern smart cards usually incorporate a java
virtual machine that is capable of interpreting multiple applets present
in the smart card's Non-Volatile Memory (NVM). There are therefore
two security considerations that need to be taken into account when
implementing a java virtual machine: secure bytecode interpretation and
secure resource partitioning. Bytecode needs to be implemented such that
the security of the smart card cannot be compromised by a malicious
applet writer, and resources managed such that one applet cannot access
the secrets of another applet. This is a relatively new problem in smart
card operating system design, as previously it was not possible to load
arbitrary applications into smart cards.

A secure smart card needs to include all of the above security considerations. There are also other security aspects that need to be taken into account
to have a totally secure solution. There are hardware features that are used to
prevent reverse engineering of a chip, its operating system, and any secret keys
held within. A discussion of this topic is beyond the scope of this chapter but
needs to be taken into account when choosing a chip for use as a smart card.

This can be mitigated by including more stringent software countermeasures to achieve a secure solution when inadequate hardware security is present.

The specific case of side channel and fault attacks applied to smart cards is covered in Section 8.2. This is followed by a discussion of multi-application platform security in Section 8.3. In Section 8.4 a case study is given detailing the changes in security in the European mobile telecommunications standards in response to various smart card based attacks. This is followed by a summary in Section 8.5.

8.2 Smart Card Specific Attacks

Sensitive systems that are based on smart cards use protocols and algorithms that have usually been subjected to rigorous analysis by the cryptographic community. Attackers have therefore sought other means to circumvent the security of protocols and algorithms used in smart card based systems. As smart cards are small, portable devices they can easily be put in a situation where their behaviour can be observed and analysed. The simplest form of analysis of this type is intercepting all the communication between a smart card and its reader. Tools are readily available on the Internet [62] that allow the commands to and from a smart card to be logged. This problem is relatively easy to solve with secure sessions, as proposed in [31], but highlights the ease of man-in-the-middle type attacks against smart cards.

More complex attacks can be realised by monitoring information that leaks naturally during a smart cards processing of information, referred to as side channel attacks. Smart cards can also be attacked by inducing a fault during their normal processing to change the chips behaviour. Comparing a faulty result with a standard response can then be used to make deductions about secrets held by a smart card. These attacks are referred to as fault attacks. The following sections describe these two classes of attack in more detail.

8.2.1 Side Channel Attacks

The first example of a side channel attack was proposed in [42]. This involved observing the differences in the amount of time required to calculate a RSA signature for different messages to derive the secret key. This attack was conducted against a PC implementation but a similar analysis could be applied to smart card implementations. It would be expected to be more efficient against a smart card as more precise timings can be achieved with an oscilloscope or proprietary readers. An example of equipment capable of acquiring this sort of information is shown in Figure 8.1. The I/O trace can be seen on the oscilloscope as the yellow trace that will allow the exact amount of clock cycles taken by a given command can be determined (the smart card being analysed is in the reader in the top left corner of the image). For this reason, the amount of time taken for a command is often constant or, if variable, not related to any secret information.

The most common form of side channel attack is the analysis of the power consumption, originally proposed in [43]. This is measure by placing a resistor in series with a smart card, between the card and earth. The potential difference across this resistor is measured with an oscilloscope. This gives a measurement of current in the circuit between the chip and the resistor. The equipment shown in Figure 8.1 is capable of taking these sort of measurements. The current at each point in time during a command is shown as the blue trace on the oscilloscope. An enlargement of this trace is shown on the computer screen, where the minimum, average and maximum of the points represented by one pixel can be seen. This is referred to as the power consumption in the literature, and the attacks based on these measurements are referred to as power analysis attacks. There are two main types of power attack; these are simple power analysis (SPA) and differential power analysis (DPA).

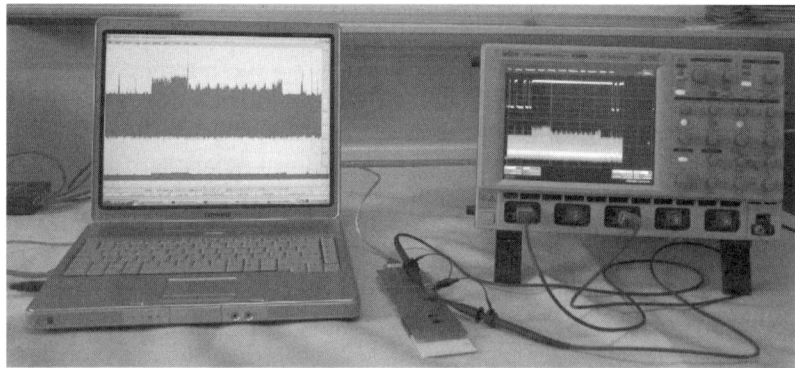

Fig. 8.1. Data Acquisition Tools.

Simple Power Analysis.

This is the analysis of one, or the comparison of a few, power consumption traces. An attacker will look for repetitive patterns and distinctive events to reverse engineer the algorithm being used and to try and derive any secrets being manipulated. Figure 8.2 shows a power consumption trace taken during the execution of an AES implementation, using equipment similar to that shown in Figure 8.1, where the 10 rounds of the DES algorithm can be observed. A pattern that repeats 9 times can be seen, with a much smaller tenth pattern, that correspond to the rounds of an AES.

Much smaller differences can be seen in the power consumption. By zooming in on specific areas the difference of 1 or more clock cycles can be used

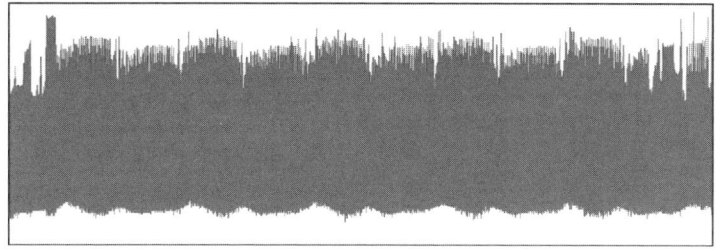

Fig. 8.2. Power consumption during the execution of an AES implementation.

to identify secret information. For example, A compiler can implement the bitwise permutations used in DES in an insecure manner. An example of an algorithm that could be produced is given in Algorithm 8.1, where each bit in buffer X (a buffer of n bits, where $(x_0, x_1, \ldots, x_{n-1})_2$ is the binary representation) is tested to determine whether a bit in buffer Y should be set.

Algorithm 8.1 Unsecure bitwise permutation function

Input: $X = (x_0, x_1, \ldots, x_{n-1})_2$, $P[\cdot]$ containing the indexes of the permutation
Output: $Y = (y_0, y_1, \ldots, y_{n-1})_2$
1. $Y := 0$;
2.   for $i = 0$ to $n - 1$
3.   if $(x_{P[i]} = 1)$ then $y_i := 1$;
4.   return Y;
end.

This is not a secure implementation due to the conditional test present in step 3. The setting of one bit in buffer Y will take a certain amount of time when the tested bit in buffer X is equal to 1. If a power consumption trace for this function is compared to a trace taken during the manipulation of a known key, for example all zeros, the point where the two traces differ will reveal the first manipulated bit that is not equal to the known key. In the case where the known key is all zeros, the power consumptions diverge where the manipulated bit of the unknown key is equal to one. All previous bits were therefore equal to 0. The power consumption trace can then be shifted to take into account the time difference due to this bit and the process repeated for the rest of the key.

This problem can be avoided by implementing the bitwise permutation as shown in Algorithm 8.2, where each bit is taken separately for buffer X and assigned to buffer Y. This is usually more time consuming than Algorithm 8.1 as each bit needs to taken from the relevant machine word.

It is not always possible to change an algorithm in this manner. For example, the modular exponentiation used in RSA is exceedingly difficult to secure against this type of attack, as branches in the algorithm cannot be avoided.

Algorithm 8.2 Secure bitwise permutation function

Input: $X = (x_0, x_1, \ldots, x_{n-1})_2$, $P[\,\cdot\,]$ containing the indexes of the permutation
Output: $Y = (y_0, y_1, \ldots, y_{n-1})_2$
1. for $i = 0$ to $n - 1$
2. $y_i := x_{P[i]}$;
3. return Y;
end.

An example of a power consumption trace that shows the square-and-multiply algorithm is given in [40].

Differential Power Analysis.

In this type of attack the smart card is forced to run the same command numerous times with different inputs. For each input the corresponding power consumption is captured and an analysis is performed on the data gathered to make inferences on the secret information held inside the card. These techniques require many more acquisitions than with simple power analysis and a certain amount of treatment *a posteriori*.

Differential power attacks [43] are based on the relationship between the current consumption of the hamming weight of the data manipulated at a particular point in time. The variation in current consumption is extremely small and acquisitions cannot be interpreted individually as the information is very small when compared to the usual amount of noise in a given acquisition. These differences can be seen in Figure 8.3, and can be amplified by using DPA.

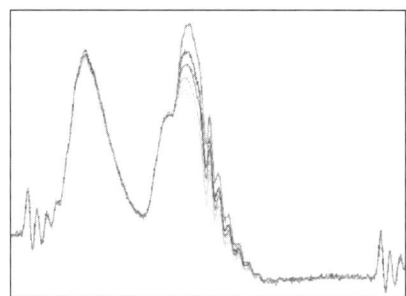

Fig. 8.3. Superimposed acquisitions of one clock cycle showing the data dependence of the power consumption.

If two average curves are produced (one where the bit is equal to one and the other where it is not), the effect of the modified bit on the current consumption becomes visible in the difference between the two averages, an example of this is shown in Figure 8.4. A peak in this waveform will occur at the moment the bit used to divide the acquisitions into two sets was manipulated. This occurs because the rest of the data can be assumed to be random and will therefore provide the same average. Figure 8.4 shows a sample DPA trace showing five different points in time where the bit used to divide the acquisitions is manipulated.

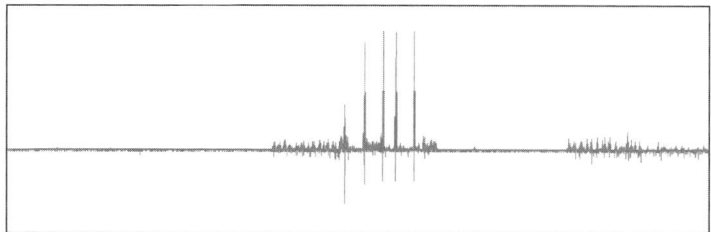

Fig. 8.4. A DPA trace.

This can be used to break a secret key algorithm if an attacker forms a hypothesis about the exit of a given function e.g. an S-box. This technique can be used to verify the hypothesis as peaks, as shown in Figure 8.4, will be present. This would appear to be analogous to an exhaustive search but the output of 1 S-box in DES depends on 6 bits of the secret key. An attacker therefore only needs to test hypotheses on these 6 bits to determine what they are. The process could then be repeated for each S-box to derive the entire key.

A method of improving this method to reduce the existence of false positives is given in [14].

Countermeasures.

There are several different countermeasures for protecting algorithms against attacks based on the power consumption, although some of the countermeasures merely complicate the attack process. The combination of all the countermeasures below makes attacking a secure algorithm exceedingly difficult, as each countermeasure needs to be taken into account.

Constant Execution: As described above, the time taken by an algorithm should remain constant, so that no deductions on secret information can be made. This extends to individual processes being executed by a smart card. If a process takes different lengths of time depending on some secret information and the difference in time is made up by a dummy function, there is a good chance that this will be visible in the power consumption as detailed in Section 8.2.1. It is therefore important that an algorithm is written so that the same code is executed for all the possible values that the secret information being manipulated can take.

Random Delays can be inserted at different points in the algorithm being executed i.e. a dummy function that takes a random amount of time to execute can be called. The algorithm can no longer be said to comply with the constant execution criteria given above, but any variation is completely independent to any secret information. This does not provide a countermeasure, but creates an extra step for an attacker. In order to conduct any power analysis an attacker needs to synchronise the power consumption acquisitions *a posteriori*. The effect of conducting statistical power analysis attacks in the presence of random delays is detailed in [18].

Randomisation (or data whitening) is where the data is manipulated in such a way that the value present in memory is always masked with the same random. This randomisation remains constant for one execution, but will vary from one acquisition to another. This mask is then removed at the end of the algorithm to produce the ciphertext. Some ideas for building countermeasures were proposed in [15], and an example of this sort of implementation can be found in [3].

The size of the random is generally limited as S-boxes need to be randomised before the execution of the random so that the input and output values of the S-box leak no information. This is done using an algorithm such as Algorithm 8.3. As shown the random used for masking the input data can be no larger than n, and the random used for the output value can be no larger that x.

Algorithm 8.3 Randomising S-box Values

Input: $S = (s_0, s_1, s_2, \ldots, s_n)_x$ containing the S-box, $\mathbf{R}$ a random $\in [0, n]$, and r a random $\in [0, x)$

Output: $RS = (rs_0, rs_1, rs_2, \ldots, rs_n)_x$ containing the randomised S-box

1.   for $i = 0$ to n
2. $rs_i := s_{(i \oplus \mathbf{R})} \oplus r$;
3.   return RS;
end.

In the case of AES both $\mathbf{R}$ and r will be one byte, which means that the random mask during the calculation is likely to be one byte. There are theoretical attacks against this protection method [52] but an actual implementation has yet to be published.

This is not possible in the case of RSA, where the calculation methods do not facilitate the method described above. A method for randomising the calculation of RSA is given in [40], where the signature generation (i.e. $s = m^d \pmod{n}$ where $n = p \times q$) can be replaced by:

$$s = \left((m + r_1 n)^{d + r_2 \phi(n)} \pmod{r_3 n} \right) \pmod{n} \tag{8.1}$$

where $\phi(\cdot)$ is Euler's totient function and, r_1, r_2 and r_3 are small random values. This does not provide a totally secure algorithm as the modular exponentiation itself also has to be secured against SPA attacks. A discussion of these algorithms is given in [17].

Randomised Execution is the manipulation of data in a random order so that an attacker does not know what is being manipulated. In the case of Algorithm 8.1 an attacker would not know which bit is being manipulated at any given point in time. Given the $n!$ possible combinations, an attack by simple power analysis would be extremely difficult.

This also inhibits any statistical analysis of the power consumption, as this relies on the same unknown variable being treated at the same point in time. As an attacker cannot know the order in which the data has been treated, this provides an extremely efficient countermeasure when combined with randomisation. An example of this technique applied to DES is described in [51].

8.2.2 Fault Attacks

The first example of a theoretical fault attack was presented in 1997 [12] as a method of injecting faults in to RSA signature generation when using the Chinese Remainder Theorem. This was followed by several other attacks on public key algorithms [5, 41] and an equivalent for private key algorithms was proposed in [9], usually based on the effect of a 1 bit error during the computation of the algorithm under study. As no implementations were forthcoming interest in this type of attack waned. One of the first publications describing an implementation of this type of attack was presented in 2002 [4], and described an implementation of the attack against the RSA signature scheme and some countermeasures. This revitalised interest in this type of attack, which is now an important aspect of smart card security.

Modelling the Effect of a Fault.

There are several known mechanisms for injecting faults into microcontrollers. These include variations in the power supply to create a glitch or spike [11], white light [63], laser light [6] and eddy currents [60]. The models for the faults that can be created by these effects can be summarised as follows:

Data randomisation: the adversary could change an arbitrary amount of data to a random value. However, the adversary does not control the random value and the new value of the data is unknown to the adversary.

Resetting Data: the adversary could force the data to the blank state, i.e., reset a given byte, or bytes, of data back to 0x00 or 0xFF, depending on the logical representation.

Modifying opcodes: the adversary could change the instructions executed by the chip's CPU. This will often have the same effect as the previous two types of attack. Additional effects could include removal of functions or the breaking of loops. The previous two models are algorithm dependent, whereas the changing of opcodes is implementation dependent.

These three types of attack cover everything that an attacker could hope to do to an algorithm. In most cases it is not usually possible for an attacker to create all of these possible faults. Nevertheless, it is important that algorithms are able to tolerate all types of fault, as the fault injection methods that may be realisable on a given platform are unpredictable. While an attacker might only ever have a subset of the above attacks available, if that attack is not taken into account it may have catastrophic consequences the security of a protocol or algorithm.

All of these models assume that a fault injected into a chip will change a machine word rather than a single bit. This model was based on studies of faults caused by the effects of the radiation present in the upper atmosphere [75], which is important in the design of vehicles designed to travel in the upper atmosphere or space. The effects of radiation are random, whereas an injected fault is intentional and will target a given point in time.

There have not been any publications that detail practical implementations of attacks that exploit the change of 1 bit of information in a chip. It is possible to use a lasers simulate the effect of radiation in a circuit [38], but smart cards will generally use scrambled/randomised layouts so the possibilities are limited. More success has been achieved by targeting larger areas to provoke a large fault.

Injecting Faults in Algorithms.

In this section two of the most widely known attacks will be described. The first fault attack published in [12] and implemented in [4], which is detailed below:

If we assume that the calculation of an RSA signature i.e. $s = m^d \pmod{n}$, where $n = p \times q$, the two primes upon which the security of RSA is based. If this is calculated using the Chinese remainder theorem the following values are calculated,

$$
\begin{aligned}
s_p &= m^{(d \,(\mathrm{mod}\ p-1))} \pmod{p} \\
s_q &= m^{(d \,(\mathrm{mod}\ q-1))} \pmod{q}
\end{aligned}
\tag{8.2}
$$

which can be combined to form the RSA signature s using the formula $s = a \times s_p + b \times s_q \pmod{N}$, where:

$$
\begin{cases} a \equiv 1 \pmod{p} \\ a \equiv 0 \pmod{q} \end{cases} \text{and} \begin{cases} b \equiv 0 \pmod{p} \\ b \equiv 1 \pmod{q} \end{cases}
$$

This can be calculated using the following formula:

$$
s = s_q + \left((s_p - s_q) \times q^{-1} \pmod{p}\right) \times q
\tag{8.3}
$$

If this is calculated correctly, and then recalculated but with a fault injected during the computation of s_p or s_q, information can be derived on one of the primes used to create n.

If s_q is changed to s'_q, then the signature generated will be of the form $s' = a \times s_p + b \times s'_q \pmod{n}$. The difference between s and s' gives:

$$
\begin{aligned}
\Delta &\equiv s - s' \\
&\equiv (a \times s_p + b \times s_q) - (a \times s_p + b \times s'_q) \\
&\equiv b(s_q - s'_q) \pmod{n}
\end{aligned}
\tag{8.4}
$$

As $b \equiv 0 \pmod{p}$ and $b \equiv 1 \pmod{q}$ it follows that $\Delta \equiv 0 \pmod{p}$ (but $\Delta \not\equiv 0 \pmod{q}$) meaning that Δ is a multiple of p (but not of q). Hence, a GCD calculation gives the secret factors of n, i.e. $p = \gcd(\Delta \pmod{n}, n)$ and $q = n/p$.

This attack could potentially be achieved with any of the fault models given above. This is because any fault in one of the modular exponentiations given in Equations 8.2 will be enough to conduct the attack. This includes modifying the variable entering the equation before the start of the computation of the modular exponentiation.

Another attack that was proposed shortly after the RSA attack was published targeting DES [9]. This assumed a 1-bit fault injected into the last few of rounds of a DES implementation. This was generalised to allow for a larger fault and became a frequently cited attack within the smart card industry. The most popular form of this attack is described in [6], and is repeated here.

DES can be considered as a transformation of two 32 bit variables (L_0, R_0), i.e. the message, though sixteen iterations of the function as shown in Figure 8.5 to produce the ciphertext (L_{16}, R_{16}). The Expansion and P permutations are bitwise permutations, and are generally not considered when studying DES. This means that the round function can be simplified to:

$$
\begin{aligned}
R_n &= S(R_{n-1} \oplus K_n) \oplus L_{n-1} \\
L_n &= R_{n-1}
\end{aligned}
\tag{8.5}
$$

where $S(\cdot)$ is the S-box function. The existence of the initial and final permutation is also ignored as they do not contribute to the security of the algorithm.

The last round, as described above, can therefore be expressed in the following manner:

$$
\begin{aligned}
R_{16} &= S(R_{15} \oplus K_{16}) \oplus L_{15} \\
&= S(L_{16} \oplus K_{16}) \oplus L_{15}
\end{aligned}
\tag{8.6}
$$

If a fault occurs during the execution of the fifteenth round, i.e. R_{15} is randomised by a fault to become R'_{15}, then:

$$
\begin{aligned}
R'_{16} &= S(R'_{15} \oplus K_{16}) \oplus L_{15} \\
&= S(L'_{16} \oplus K_{16}) \oplus L_{15}
\end{aligned}
\tag{8.7}
$$

If we XOR R_{16} and R'_{16} we get:

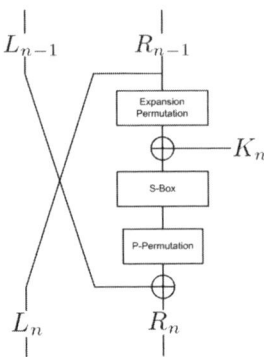

Fig. 8.5. The DES round function for round n.

$$
\begin{aligned}
R_{16} \oplus R'_{16} &= S(R_{15} \oplus K_{16}) \oplus L_{15} \oplus S(R'_{15} \oplus K_{16}) \oplus L_{15} \\
&= S(R_{15} \oplus K_{16}) \oplus S(R'_{15} \oplus K_{16}) \\
&= S(L_{16} \oplus K_{16}) \oplus S(L'_{16} \oplus K_{16})
\end{aligned}
\tag{8.8}
$$

This provides an equation where only the last subkey, K_{16}, is unknown. All of the other variables are visible in the ciphertext. This equation holds for each S-box in the last round, which means it is possible to search for key hypotheses in sets of six bits. All 64 possible key values corresponding to the XOR just before each S-box are exhausted to generate a list of possible key values for these key bits. After this, all the possible combinations of the hypotheses can be searched though with the extra 8 key bits that are not included in the key to find the entire key.

If R'_{15} becomes random then the expected number of hypotheses that are generated can be predicted. For a given input and output difference there are certain number of values that could create the pair of differences, as described in [8]. The expected number of hypotheses for the last subkey will be around 2^{24}, giving an overall expected keyspace of 2^{32}.

If the attack is repeated to acquire two faulty ciphertexts the intersection of the two keyspaces can be taken. This will greatly reduce the keyspace that will need to be searched through to derive the key. It would be expected that two faulty ciphertexts with the properties described above would give around 2^6 hypotheses for the last subkey, leading to an exhaustive search of around 2^{14} for the entire key.

An implementation of this is described in [30] where two faulty ciphertexts are acquired, leading to a small exhaustive search to find the DES key used. Again, it would be expected that any of the fault models given would enable this attack to be conducted. As any fault during the calculation of R_{15} will give a ciphertexts with the correct properties.

This attack is also extended in [30] to allow for any faults from the eleventh round onwards, the details of which are beyond the scope of this article.

Another attack [32] attempts an attack of a similar nature but using faults at the beginning of the DES algorithm.

This gives two different attacks against commonly used algorithms. There are a plethora of other attacks that target other algorithms, but they cannot be listed here. These two attacks were chosen to represent the fault attacks that have been proposed and implemented.

Countermeasures.

There are various types of countermeasure that can be implemented to defend against fault attacks. These are usually based on existing techniques used for integrity purposes. It would normally be expected that anomaly sensors would be implemented in a smart card that would detect an attempted fault attack. However, this cannot be relied upon as a new fault injection technique may be able to circumvent these.

A detailed list of the possible fault resistant hardware implementations is given in [6]. Some of the software countermeasures that can be used are listed below.

Checksums can be used to verify the integrity of data at all times. This is especially important when data is being moved from one memory area to another. For example, a fault injected in an RSA variable can compromise the secret key, as described above.

Variable redundancy is the reproduction of a variable in memory and functions are performed on each variable independently. The result will be known to be correct if both answers are identical.

Execution redundancy is the repetition of the same function, part of the function or it's inverse. In the case of the DES algorithm it would be prudent to repeat the first 3 rounds to protect against the attack described in [32], and the last 5 rounds to protect against the attack described in [9, 30]. This greatly increases the execution time but is less costly than repeating the entire algorithm.

In the case of RSA the simplest solution to protect the signature generation function is to verify the result with the public key. This is efficient as signature verification is extremely fast when compared to signature generation. In some standards it is not always possible to have access to the public key, countermeasures have therefore been proposed that verify certain conditions after signature generation has taken place. An example of this type of countermeasure, and an account of previous methods, is given in [16].

Execution Randomisation: If all the functions are conducted in a random order it is not be possible to determine exactly where a fault needs to be injected. The presents a similar problem to that described in Section 8.2.1, as an attacker is unsure of what function is being attacked. However, this merely slows an attacker as an attack can be repeated until successful. An example of this type of countermeasure on the context of fault attacks is given in [55].

Ratification counters and baits: A countermeasure described in [6] involves including small functions in sensitive code that perform a calculation and then verify the result. When an incorrect result is detected a fault in known to have been injected. The reaction to such an event would be to decrement a counter. When this counter reaches 0 the smart card would then cease to function. When combined with random delays, as described in Section 8.2.1, this can be a very effective countermeasure.

In order to achieve a secure implementation the above countermeasures would need to be combined with those presented in Section 8.2.1. This implies a significant overhead when implementing cryptographic algorithms for smart cards but is necessary to defend against modern attack techniques.

8.3 Smart Card Platform Security

The aim of this section is to provide an overview of the issues, apart from the underlying hardware functionality, affecting the overall concept of smart card security. Therefore, it will explore the functionality offered by multi-application smart card platforms and it will also highlight the main security concerns behind them.

8.3.1 The Evolution of Smart Card Platforms

The evolution of smart card operating systems (SCOS) and platforms is very closely coupled with the various improvements of the underlying smart card hardware. A typical smart card of the mid 90's had approximately 1-3 Kbytes of ROM, less than 128 bytes of RAM and 3-6 Kbytes of Electrically Erasable Programmable Read Only Memory (EEPROM). Within this very restrictive processing environment the role of the smart card operating system was naturally limited.

In most cases, the role of these early smart card operating systems (i.e. first generation) was to mainly offer the basic functionality defined within the various smart card standards [35–37] and, more specifically, to present a protected file system. This was because as smart card microprocessors were mainly perceived as secure storage devices. The main drawback of these operating systems was that the smart card applications developers had very little flexibility in terms of developing a smart card application, as the functionality offered was mostly implemented in the Read Only Memory (ROM) of the card. Although, some of these operating systems [29, 56] claimed that they offered multi-application smart card functionality, the reality was different.

The next generation (i.e. second generation) of smart card operating systems, e.g. Multos [47] emerged between 1995 and 1999 with the introduction of more powerful microprocessors consisting of 6-8 Kbytes of ROM, 128 bytes of RAM and 6-12 Kbytes of EEPROM. The smart card microprocessors were capable of hosting what it often referred to as the "monolithic" [34, 48] smart card operating systems. The main characteristic of the "monolithic" operating system was that its functionality was enhanced compared to the previous generation due to a greater underlying set of commands. Smart card applications and operating systems were still very closely coupled together within the ROM memory of the card and, although they were more advanced, portability was a major concern.

The third generation of smart card operating systems came into existence from approximately 1999 onwards. The market requirements were more mature in terms of application interoperability, hardware independence, and post issuance capabilities. Among the main characteristics of these operating systems was that applications and operating system functionality were written to the EEPROM and ROM respectively, see Figure 8.6. Moreover, the presence of a Hardware Abstraction Layer (HAL) ensured that application portability should no longer be an issue.

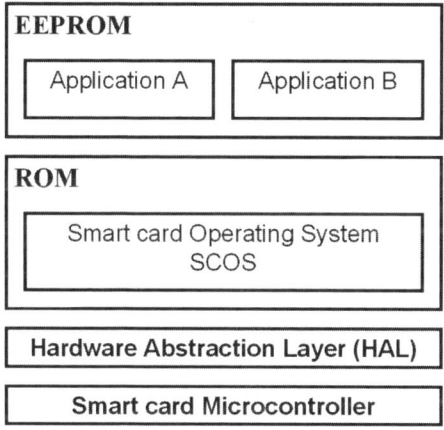

Fig. 8.6. The typical architecture of a third generation SCOS.

8.3.2 The Different Multi-application smart card Platforms

Among the main characteristics of secure Multi-application smart card platforms was application interoperability and inter application isolation. Smart card issuers would like to be able to develop smart card applications that would be interoperable irrespectively of the underlying smart card hardware. Similarly, they were looking for an underlying platform that would be able securely isolate one application from another by performing efficient memory management.

The two principal Multi-application smart card standards are Java Card [21, 67–69] and MULTOS [47]. However, there are two other technologies namely the Smartcard.NET [39, 64] of Hive-Minded and Multi-application BasicCard ZC6.5 [74] of ZeitControl. The main Multi-application smart card platforms are presented below, excluding the Java Card that will be discussed in the next section.

WfSC.

Around 1998, Microsoft launched a new business proposition under the name of Windows for Smart Cards (WfSC). The whole concept was defined in [45], where the underlying platform was capable of performing dynamic application management and the applications were developed in Visual Basic (VB) or C. The philosophy of the proposed architecture [61] was that smart card programming should become part of mainstream programming rather than to force programmers to acquire new competences and new models of programming. In fact, under this framework a WfSC project was simply another type of Visual BASIC or Visual C++ project. A couple of years afterwards, Microsoft dismantled the WfSC team as it became evident that it was no longer directly supporting it. The most notable observation about WfSC was that the overall security functionality was reviewed and guaranteed by Microsoft, which, along with the fact that it was a "closed" system, created barriers in terms of the technology adoption.

Multos.

The Multos [47] smart card operating system was specifically designed for smart card microprocessors. Security was taken into serious consideration right from its conceptualisation. An industrial consortium, named MAOSCO, was created in order to promote MULTOS as a multi-application operating system for smart cards, to manage the specifications MULTOS, and to provide the licences and certifications of MULTOS services. Multos is an operating system that also offers dynamic application management and secure application isolation right at the operating system kernel. Some of the key elements of the MULTOS platform are:

- a highly protected architecture;
- the possibility of having several applications on the same card;
- independence of the applications compared to the underlying platform;
- compatibility with industrial standards such ISO 7816 and EMV [12].

Applications can be developed in a language optimised for smart cards: MEL (Multos Executable Language) based on the Pascal P-codes and the virtual machine described in [72]. From a security point of view, Multos achieved one of the highest certified security levels (i.e. EL6 of ITSEC [44] which corresponds to EAL7 of Common Criteria [20] security evaluations) ever achieved by a commercial product. Although this provided some security reassurance it also increased the overall cost of the platform.

The ZeitControl BasicCard.

The concept of BasicCard came into existence around 1996 with the first card being released in 2004, i.e. MultiApplication BasicCard ZC6.5 [74]. The BasicCard was programmed in ZeitControl BASIC and it also contained a virtual machine. The main difference to the other card was that it was the only one capable of supporting floating numbers [61]. Among the main advantage of the Basic card was that it was a low cost card and it offered a relatively good starting point for smart card programmers

8.3.3 Java Card

Java was among the very first programming languages that took security into account and, at the same time, enjoyed wider acceptance. The general Java language can be described as an "object-oriented (with single inheritance), statically typed, multithreaded, dynamically linked and has automatic garbage collection [where] application developers can use a number of Application Programmers Interfaces (API) when developing new applications" [27]. It is generally accepted that the Java programming language provides a secure model that prevents programs from gaining unauthorised access to sensitive information. At the same time, since Java is based on a runtime byte code interpreter, the portability issue is successfully addressed.

Although Java Card may be perceived as another flavour of traditional Java, the reality is different. The first Java Card API (version 1.0) [65] was released in October 1996 as the initial attempt to bring the benefits of the Java language into the smart card world. Since then, with major enhancements taking place around 1999 with the release of the Java Card API version 2.0 [66], further versions were released incorporating major improvements. Java Card preserves the object-oriented scope of the Java programming language, but it is essentially a subset of normal Java and does not support such things as: Dynamic class loading, security manager, object cloning, threads, and large primitive data types.

The internal architecture of the Java Card specification is illustrated on the right hand side of Figure 8.7. At the bottom layer of the smart card architecture is the hardware (i.e. the actual smart card microprocessor). In the next layer, we encounter the smart card operating system (SCOS). On top of the SCOS, we have the Java Card Virtual Machine (VM). Both the Java Card VM and the SCOS are written in the native language of the microprocessor. This abstraction hides the manufacturer's proprietary technology with a common language interface.

The steps for creating a Java application (the so called applet), download it and execute it on the smart card are the following: First of all, the application programmer must take into account the Java Card APIs and develop an application by using a standard Java development environment. Since there is rarely a byte code verifier in the card, the newly created Java code classes should be verified externally. This implies that the Java Card VM may rely on a digital signature before accepting code from an external source. An external converter is also utilised in order to reduce the size of the actual application. As soon as the Java classes are verified, the application code is ready for loading onto the card via the external application loader. The whole procedure is summarised on the left hand side of Figure 8.7.

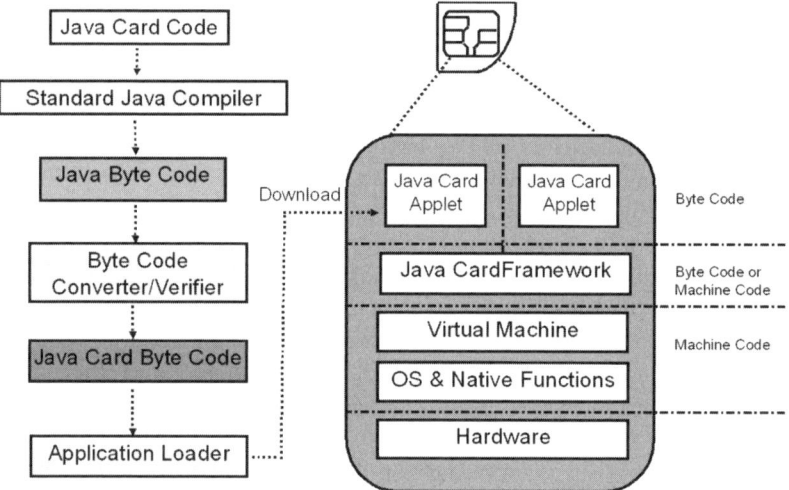

Fig. 8.7. The Java Card application development cycle and the Java Card architecture.

The Java Card platform security feature are inherited from the Java language and several enhancements are also provided through the Java Card framework and run-time environment. The fact that certain Java functionality was removed from the Java Card framework in order to ensure that the technology is ported to smart cards introduced some potential security problems.

The discussion around the security issues introduced is very closely coupled with the security functionality offered by the underlying platform. Therefore, in the next few sections the reader will be presented with the differences and additional functionality introduced by the Java Card platform in relation to the most notable Java Card attacks and vulnerabilities.

8.3.4 Java Card Security

Although there are many attacks directed against the Java Card platform, it is not possible to provide their details here, as they have been explained extensively in the academic literature. However, our discussion on the Java Card risks will also include particular references to the differences with traditional Java that remove certain security problems.

The lack of dynamic class loading within Java Card makes the overall static code verification process simpler. However, it is not possible to guarantee the type safety of classes that are about to be loaded. Furthermore, the fact that only a single Java Card application can be active and executing at any point in time (single threading) further simplifies the overall Java Card security model. Java Card has been extensively criticised in that the programming functionality is seriously restricted within the Java Card API, which can be particularly restraining when new powerful methods need to be implemented, e.g. cryptographic operations. However, the existence of a mechanism that will bypass the Java language syntax and type safety features could allow a programmer to execute arbitrary code, which would have been a major concern for smart card vendors and issuers as it could create serious access violations. Among the most notable properties related to the Java Card security model are the following [49, 73]:

- Garbage collection
- CAP verification
- Applet loading
- The firewall mechanism
- Transaction atomicity
- Native methods
- Exceptions

The lack of a garbage collection mechanism, which is an optional feature, can be a major concern. The idea in normal Java is that unused space is collected and reused. This process is achieved though the existence of a complex mechanism. The so-called "garbage collector" is a relatively large in terms of code size and it was therefore not possible to be defined as a mandatory object within the Java Card specifications. This opens the opportunity to create a malicious Java Card applet that leaks memory resulting in a denial-of-service attack.

The CAP file is a representation of the original Java binary file, modified due to the limited resources available to a given smart card. It is stated in [73] that since the CAP file is restricted in size, it has also lost some of the inherent Java language security features. It is considered to be possible to modify the CAP file and change it in a way that will be able to threaten the security of the underlying platform. Therefore, it is very important to be able to verify that the CAP file adheres to the Java Card security rules. Although Java Card developers initially believed that it was impossible to include a Java

Card verifier on a Java card, due to the various resource constraints. Over the last few years on-card Java verifiers have been developed [19,46,59]. Some of these solutions "allow to bypass the signature step of the application without jeopardising the card security". However, the existence of a defensive VM that will dynamically execute the byte codes may be preferable to a stand alone verifier will statically check during the load time [61].

Application downloads, especially at post issuance, lies within the core concept of a multi-application smart card. If applications can be added deleted or modified at post issuance, then the issuers are presented with a relatively powerful and flexible platform that will meet their future requirements. However, the existence of an application downloading mechanism within the Java Card API is not defined. The Global Platform [31] specification comes into play, in order to handle the security sensitive operation of post issuance application downloading. The details of the Global Platform specification are widely known and beyond the scope of this analysis. The Global Platform specification is an important element of Java Card security as it takes care of further security risks. For example, it strengthens the applet firewall (to be explained below) by the existence of a card manager. The Global Platform card manager is responsible, among others, for ensuring that only trusted code (bearing the appropriate credentials) can be downloaded to the card. Therefore, the existence of trusted applications (that have been previously verified) facilitates the existence of an elaborate application sandbox. This in turn requires the existence of thorough application testing and verification process.

The Java Card firewall mechanism is responsible for isolating applications from each other. It ensures that no application is allowed to access another application or its data in an authorised way. The firewall is considered of paramount importance in a multi-application smart card environment, as there is always the risk of inter-application attacks. By default the Java Card [73] applications are only allowed to access data objects that they own. The mapping between what is owned by an application is done during the application downloading and installation process. It is obvious that this process requires the existence of memory protection mechanisms that will perform the necessary checks. The role of the firewall is also to control authorised data sharing. In terms of firewall security there are plenty of attacks that have been described in the literature. "Two attacks against the firewall mechanism (AID) impersonation and illegal reference casting [54] that provides access to all interface methods of a class, along with other attacks originating from problems in the specifications [73]" [22]. A further type of attacks is based on "type confusion" and it is further described in [7].

Transaction atomicity is a further important concept that aims to improve the overall stability of the Java Card platform. Smart cards are often operating in environments in which a user can remove a smart card from its reader at any point in time. In all cases, smart cards, and particularly Java cards, should be in a position to recover when the normal operation is disrupted, e.g.

when power is lost. Transaction atomicity guarantees that any updates to a single persistent object or class will be atomic (i.e. either not performed at all or fully performed). Automatic transaction integrity describes how the virtual machine behaves when power is lost during the update of a field in an object. On the other hand, block transaction integrity describes the virtual machine behaviour when the application programmer identified a specific part of his code that needs to be executed in one piece. For example consider a code example presented in [70] (*CommitBuffer.begin*(), ATC++, ATC++, *CommitBuffer.commit*()). The ATC is a simple Application Transaction Counter. If such a block is defined (*CommitBuffer.begin*()) and not ended with (*CommitBuffer.commit*()) an automatic rollback is performed at the end of the application execution or when card is powered back on.

Native methods are not allowed by the Java Card specifications. However, Java Card application developers would love to be able to access specific calls that will allow them to perform low level tasks, e.g. by obtaining direct access to the any cryptographic coprocessors. As mentioned before, implementing some of the Java Card API functionality requires using native methods. The existence of native methods will completely violate the language based encapsulation model [50] and this one of the main reasons that smart card vendors decided not to offer such functionality. The existence of native methods would also compromise portability.

Exceptions are thrown by the Java Card VM (when internal runtime problems are encountered) or they can be thrown programmatically (Checked exceptions, Unchecked exceptions). The Java Card VM catches the exceptions that are not caught by the application. The Java Card application programmers can define their own exceptions by declaring subclasses of the class Exception. The wide utilisation of exceptions is another interesting topic within Java card. It is mentioned in [49] that unhandled exceptions may cause denial of service attacks. The significance of exception handling was recognised in the Java Card API version 2.1 when the "Exception" class was redesigned.

Java Card CAP file reverse engineering is a potentially serious type attack. It requires obtaining access to a CAP file and running it through a de-compiler, to find out as much information as possible about the application. The next step requires importing certain changes and minor modifications, in order to change the behaviour of the application. However, although minor changes in the CAP file create further changes in the complete file, it is possible to perform pattern changes [23] that will reveal certain information of the application execution.

8.4 GSM and 3G Security

Smart cards play an important role in mobile telecommunications where their security functionality combined with tamper-resistance, underpins the control and viability of the industry. However this was not always the case and the use of smart cards in 2G & 3G systems (such as GSM and UMTS) is largely due to problems that occurred in early systems. The lessons learned form the evolution of the security in mobile phones have played a large part in defining the security issues discussed in Sections 8.2 and 8.3.

8.4.1 1G - TACS

Total Access Communication System (TACS) - was an analog cellular phone systems working at 900MHz and initially deployed within various European countries including UK, Italy, Austria and Spain. TACS was derived from the US cellular system known as Advanced Mobile Phone Service (AMPS). The first commercial TACS system was deployed in the UK and was put into service by 1985.

Early mobile phone systems such as TACS started life using analogue technology and handsets that did not contain smart cards. There was however a security solution embedded in the handsets and network, designed to ensure that only valid users were allowed access to communications services and that they were billed appropriately. This solution was limited to some extent by the available technology but it is probably fair to say that the technical emphasis of the day was primarily to get the radio system to work rather than on its security.

When an analogue phone was switched on, or made a call request, it transmitted (in clear) two important pieces of information; the SNB and the ESN. The SNB was the Subscriber Number (telephone number) and the ESN was the handset Electronic Serial Number. When the network received this information it checked that the transmitted ESN matched the stored value within the network for that particular SNB. In theory the ESN was fixed per phone and the SNB could only be changed by the operator, however many phones were easily tampered with, meaning that both fields could be reprogrammed to create clones.

Typically, an attacker would purchase a radio scanner to monitor the frequencies used by the analogue networks and set up near to an area where analogue phones were likely to be used e.g. motorway services or airport. ESN-SNB pairs could then be captured and this information programmed into another handset. The network would then permit the use of the handset and the charges for the calls made on the "cloned" phone would appear on the legitimate customer's bill. The end result was a major headache for the mobile operators who not only lost revenue because of cloning but also from customers making false claims of cloning to dispute their bills. By 1995 the problems in the UK had become so severe that Members of Parliament were briefed on the issues [57].

8.4.2 2G - GSM

The industry reaction was to ensure that the next generation of digital mobile communication system (GSM) being standardised by ETSI [24] would be designed to avoid clear transmissions of security keys and not be reliant on the discredited tamper-resistant capabilities of handsets. The answer was the introduction of smart cards in the form of Subscriber Identity Modules (SIMs) that are described in detail by GSM 11.11 [25]. From an access and security viewpoint the SIM had to provide integrity, authentication and help ensure confidentiality.

The identity that is important to the GSM network is the International Mobile Subscriber Identity (IMSI) stored in the SIM. It is not the users telephone number but a unique number associated with a customer account. It cannot be altered but is otherwise not very well concealed, as when a user first arrives on a network the IMSI is transmitted in clear over the radio interface. Thereafter a Temporary Mobile Subscriber Identity (TMSI) is allocated and used but this just adds a little extra inconvenience for the attacker. The IMSI can also be read by putting a SIM card in a smart card reader (providing it is not PIN-locked).

The stronger authentication security comes from an algorithm and a 128 bit secret key (Ki) that is stored in the SIM and mimicked in a network system called the Authentication Centre (AuC). Logically the AuC is part of the Home Location Register (HLR) that holds a mapping of all the IMSIs and Kis. The AuC generates the necessary challenges and expected responses to ensure that the SIM card is genuine. The algorithm is sometimes referred to as A3/8 and is in fact two algorithms, one to produce the authentication result (A3) and the other (A8) to produce a radio cipher key. A basic diagram is shown in Figure 8.8.

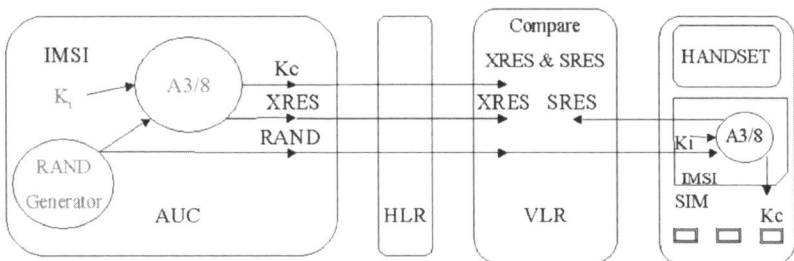

Fig. 8.8. GSM authentication.

When the user wishes to access the GSM network it is not sufficient just to present the IMSI, but also to prove that a valid SIM card is present. This works as detailed in Algorithm 8.4.

This introduction of the tamper-resistant SIM solution was a huge leap forward and overcome the problems so prevalent in the analogue phone system i.e. it prevented simple cloning and offered some confidentiality on the radio interface. History has shown that the SIM has done a pretty good job at maintaining mobile communications over a period of many years, although

---

Algorithm 8.4 Accessing the GSM network

1.   The AuC Generates a 128 bit random number (RAND).
2.   The AuC runs the algorithm which uses RAND and secret key (Ki) as inputs.
3.   The expected result is produced (XRES) plus a cipher key (Kc) for encrypting radio transmissions.
4.   RAND, XRES and Kc. The triplet is handed to a Visitor Location Register (VLR) that is in the same location area as the user.
5.   RAND is passed to the SIM via the radio network and the handset.
6.   The SIM runs its copy of the algorithm and generates a result (SRES) and a cipher key (Kc).
7.   SRES is sent back to the VLR.
8.   If SRES equals XRES the user is authenticated.
9.   Kc is then used for ciphering subsequent communications.
end.

---

cannot be claimed to be perfect and its weaknesses have been well publicised. Like all smart cards it can be subject to the security attacks described in earlier sections, but the smart card industry has been vigilant in this respect and modern cards implement robust good countermeasures and are difficult to tamper with.

However, this does not help if a poorly designed security algorithm is used that will reveal its secrets to a logical or brute force attack. There is one particular algorithm called COMP128-1 that is invariably given as an example of, not only a weak authentication algorithm, but also to illustrate the perils of "security by obscurity". Whilst the 128 bit Ki should have held off brute force attacks, the leakage of the design documentation was swiftly followed by researchers at Berkley [13] extracting a secret Ki in about 130K-160K algorithm executions. The researchers had spotted a flaw that made the algorithm vulnerable to collisions i.e. different RAND challenges giving the same output. The algorithm used multiple rounds of compression based on the Ki and RAND values, but referring to Figure 8.9 one can see that the output bytes of the second round are only dependent on 4 input bytes. The final gift to the attackers was that a collision occurring at the second round propagates to the final result. This is described in more detail within the literature [13].

The immediate response was to introduce an authentication retry count into the SIM card so that it would cease to function before the required number of executions had been performed. However, techniques were developed to attack the subsequent rounds of the compression that brought the number of required trials down to 20'000. Side channel leakage was also investigated at IBM [58] bringing the number of executions down to the range of 16 to 1000 to extract the key using a particular form of SPA.

The inescapable conclusion is that there is no reasonable excuse for a GSM network to still issue COMP128-1 SIMs. A common misunderstanding is that it is "the" GSM algorithm whereas it was really only an example made

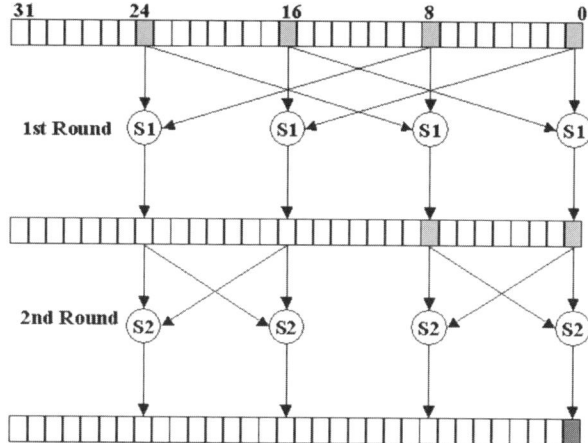

Fig. 8.9. Compression Rounds in Comp128-1.

available to members of the GSM association. Standardisation only defined the method of authentication and network operators are free to specify or develop their own algorithms. The standards also permit the use of AuCs that support multiple authentication algorithms which means that networks can introduce new algorithms whilst gradually phasing out legacy SIMs

Whilst the problems inherent to COMP128-1 were perhaps avoidable there were other security gaps due to the overall GSM security architecture. This is not to say that GSM was badly designed but just that it was a solution to the problems of the previous generation of phones, notably to stop a cloning mechanism, prevent simple radio eavesdropping and make it harder to discover the subscriber's identity. Because of this, GSM is still vulnerable to a type of attack called man-in-the-middle, which becomes possible when a false base station is used as illustrated in Figure 8.10. The attack is described is described in GSM and UMTS [33], and is possible because there is no mutual authentication and the user has no way of proving that the network or base station is legitimate.

Another vulnerability relates to the ciphering that is actually carried out in the handset using an algorithm usually called A5/1 (others are available) using the cipher key (Kc) generated by the A8 algorithm in the SIM. There are academic papers [10] describing attacks on the A5/1 algorithm but this seems not to have resulted in a practical criminal exploitation. This is partly due to the difficulty in collecting sequences of ciphertext and corresponding plaintext but more likely because there are easier ways to access the transmitted information.

Aside from the false base station method, there are other opportunities because ciphering is only applied between the user and the base station, everywhere else in the network the information is unprotected. This means that

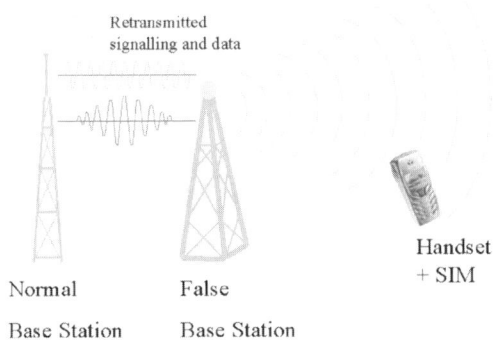

Fig. 8.10. False Base Station Attack Scenario.

a hacker needs physical access to the network but there are miles of cabling to attack as well as microwave links used between some base stations and core network equipment.

8.4.3 3G - UMTS

In designing the successor to GSM the goal was to further exploit the capabilities of the SIM (referred to as a UMTS SIM or USIM in this context) to remove security weaknesses in the GSM architecture and its algorithms, especially in the light of massive advances in computing power. This work was originally driven by ETSI but was eventually handed over to the 3GPP organisation [1]. The most important change was the introduction of mutual authentication between the user and the network based on a strong algorithm. Whereas the description of COMP128-1 was intended to be a secret, the ETSI SAGE team [26] defined a default 3G algorithm (Milenage) based on AES that has been open to expert public scrutiny. The architecture is also improved compared to GSM. The cipher key is longer and there are new integrity (IK) and anonymity (AK) keys plus a sequence number (SQN) to provide replay attack protection. The complete solution at the network end is shown in Figure 8.11 with further detail being found in [2].

The basic authentication process from the user/USIM perspective is as follows: The AuC generates an Authentication Vector (AV), this containing a secret value (MAC) that is generated with knowledge of the subscribers key (K) and the sequence number (SQN). This process is shown in Figure 8.11. To check the MAC the USIM has to follow the sequence shown in Figure 8.12. The USIM generates AK from its own K and the RAND sent from the AuC using function f5. AK is then used to recover the SQN, if this is valid the USIM will then generate the expected value of MAC called XMAC. If XMAC and MAC are identical the network is authenticated. The USIM can then

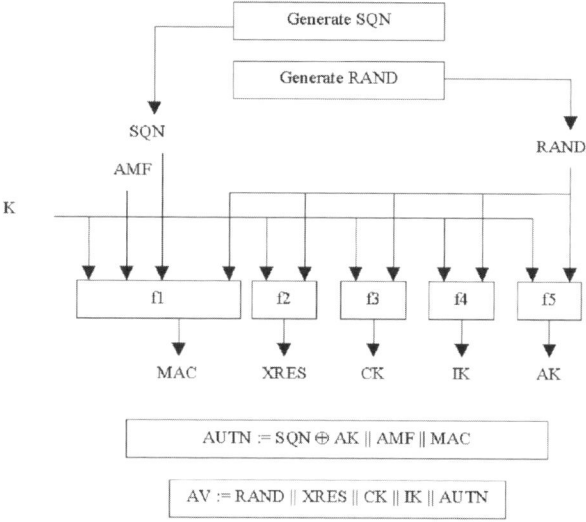

Fig. 8.11. UMTS Network Authentication Vector Generation - source [2].

calculate RES, CK and IK that are used to authenticate the USIM to the network.

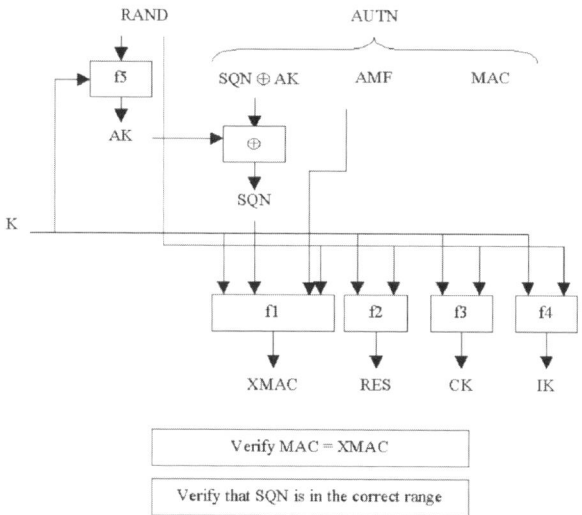

Fig. 8.12. USIM Authentication Vector Handling - source [2].

This solves many of the anticipated problems, but whilst ciphering no longer terminates at the base station and protects some signalling information, it is not end-to-end. The end-point is in fact further back in the network than GSM at a node called the Radio Network Controller (RNC) and this safeguards a lot more of the back-haul, the core interconnect remains unencrypted. Another problem is that users may dynamically switch between GSM and UMTS networks based on coverage, performance and service issues. The hand-over between the networks is regarded by some as a security weak point and researchers have sought to use combinations of GSM attacks to try and undermine UMTS [53].

The above text has introduced and compared only the most fundamental security functions of GSM and UMTS that are underpinned by the tamper-resistant functionality of SIM and USIM smart cards, and shown how the security has improved over time in response to fraud. However the (U)SIM provides a range of auxiliary security functions including PIN management, file access control, remote file and application loading and management either locally or Over The Air (OTA). There is also increasing interest in providing APIs (such as JSR177 [71]) in handheld devices so that the security functionality may be accessed and exploited to help protect handset applications and services. In fact with the introduction of Java card and Global Platform functionality the USIM in common with other UICC based smart cards, has become a powerful multi-application Platform. Whilst this brings new opportunities it also beings new challenges and possibilities for attack.

8.5 Summary

Smart card security can be analysed like many other systems as a stack of services each relying on the security of the lower levels. For example, the application developers should be able to rely on the security of the underlying platform. The platform developer relies on the security of the cryptographic algorithms and the hardware being used.

This chapter has provides an overview of the potential problems on smart card security and how these can be avoided. A case study on the European mobile telephone standards is used as a case study to show how security requirements have evolved over time due to the system being exploited.

Smart cards are often considered the magic solution to a problem; often this just removes a weak link in a protocol. The rest of the process also needs to be analysed to ensure that the whole process is secure.

Research in the domain of smart card security is a cyclic process. New attacks are developed against algorithm implementations, standards, APIs, etc. and countermeasures are proposed. The modifications are then reviewed for potential vulnerabilities and further countermeasures proposed if required. The aim of this process is to remain sufficiently ahead of what can be achieved by an individual attacker that smart cards remain secure throughout the period they are active.

References

1. 3GPP organisation. http://www.3gpp.org/
2. 3GPP organisation. 3GPP TS 33.102 3G Security; Security Architecture (Release 99) V3.13.0 (2002-12), 2002
3. M.-L. Akkar and C. Giraud. An implementation of DES and AES secure against some attacks. In Ç. K. Koç, D. Naccache, and C. Paar, editors, *Cryptogaphic Hardware and Embedded Systems — CHES 2001*, volume 2162 of *Lecture Notes in Computer Science*, pp. 309–318. Springer-Verlag, 2001
4. C. Aumüller, P. Bier, P. Hofreiter, W. Fischer, and J.-P. Seifert. Fault attacks on RSA with CRT: Concrete results and practical countermeasures. In B. S. Kaliski Jr., Ç. K. Koç, and C. Paar, editors, *Cryptogaphic Hardware and Embedded Systems — CHES 2000*, volume 2523 of *Lecture Notes in Computer Science*, pages 260–275. Springer-Verlag, 2002
5. F. Bao, R. H. Deng, Y. Han, A. Jeng, A. D. Narasimhalu and T. Ngair. Breaking Public Key Cryptosystems on Tamper Resistant Devices in the Presence of Transient Faults, the Proceedings of the *5th Workshop on Secure Protocols*, volume 1361 of *Lecture Notes in Computer Science*, Springer-Verlag, pp. 115–124, 1997
6. H. Bar-El, H. Choukri, D. Naccache, M. Tunstall, and C. Whelan. The sorcerers apprentice guide to fault attacks. *Proceedings of the IEEE: Special Issue on Cryptography and Security*, 94(2):370–382, IEEE, 2006
7. G. Betarte, E. Gimenez, B. Chetali, and C. Loiseaux. FORMAVIE: Formal Modeling and Verification of Java Card 2.1.1 Security Architecture, In Proceedings of E-Smart 2002, pp. 215–229, 2002
8. E. Biham and A. Shamir. Differential cryptanalysis of DES-like cryptosystems. In A. Menezes and S. Vanstone, editors, *Advances in Cryptology — CRYPTO '90*, volume 537 of *Lecture Notes in Computer Science*, pp. 2–21. Springer-Verlag, 1991
9. E. Biham and A. Shamir. Differential fault analysis of secret key cryptosystems. In B. S. Kaliski Jr., editor, *Advances in Cryptology — CRYPTO '97*, volume 1294 of *Lecture Notes in Computer Science*, pp. 513–525. Springer-Verlag, 1997
10. A. Biryukov, A. Shamir, and D. Wagner. Real time cryptanalysis of A5/1 on a PC, In B. Schneier, editor, *Fast Software Ecryption — FSE 2000*, volume 1978 of *Lecture Notes in Computer Science*, pp. 1–18, Springer-Verlag, 2000
11. J. Blömer and J.-P. Seifert. Fault based cryptanalysis of the advanced encryption standard (AES). In R. N. Wright, editor, *Financial Cryptography*, volume 2742 of *Lecture Notes in Computer Science*, pp. 162–181. Springer-Verlag, 2003
12. D. Boneh, R. A. DeMillo, and R. J. Lipton. On the importantce of checking computations. In W. Fumy, editor, *Advances in Cryptology — EUROCRYPT '97*, volume 1233 of *Lecture Notes in Computer Science*, pages 37–51. Springer-Verlag, 1997
13. M. Briceno, I. Goldberg, and D. Wagner. GSM Cloning. 20 April 1998. http://www.isaac.cs.berkeley.edu/isaac/gsm.html

14. E. Brier, C. Clavier and F. Olivier. Correlation power analysis with a leakage model. In M. Joye and J.-J. Quisquater, editors, *Cryptographic Hardware and Embedded Systems — CHES 2004*, volume 3156 of *Lecture Notes in Computer Science*, pp. 16–29. Springer-Verlag, 2004

15. S. Chari, C. S. Jutla, J. R. Rao, and P. Rohatgi. Towards approaches to counteract power-analysis attacks. In M. Wiener, editor, *Advances in Cryptology — CRYPTO '99*, volume 1666 of *Lecture Notes in Computer Science*, pp. 398–412, Springer-Verlag, 1999

16. M. Ciet and M. Joye. Practical fault countermeasures for chinese remaindering based RSA. In L. Breveglieri and I. Koren, editors, *Workshop on Fault Diagnosis and Tolerance in Cryptography 2005 — FDTC 2005*, pp. 124–131, 2005

17. B. Chevallier-Mames, M. Ciet, and M. Joye. Low-cost solutions for preventing simple side-channel analysis: side-channel atomicity. *IEEE Transactions on Computers*, 53(6):760–768, IEEE, 2004

18. C. Clavier, J.-S. Coron, and N. Dabbous. Differential power analysis in the presence of hardware countermeasures. In Ç. K. Koç and C. Paar, editors, *Cryptographic Hardware and Embedded Systems — CHES 2000*, volume 1965 of *Lecture Notes in Computer Science*, pp. 252–263. Springer-Verlag, 2000

19. R. Cohen. The defensive Java virtual machine specification, Technical Report, Computational Logic Inc., 1997

20. Common Criteria. `www.commoncriteria.org`

21. Z. Chen. Java Card Technology for Smart Cards : Architecture and Programmer's Guide, Addison-Wesley, 2000

22. S. Chaumette and D. Sauveron. Some Security Problems Raised by Open Multi-application Smart cards, 10th Nordic Workshop on Secure IT-systems — Nord-Sec 2005, 2005

23. S. Chaumette and D. Sauveron, An efficient and simple way to test the security of Java cards, In Proceedings of the 3rd International Workshop on Security in Information Systems — WOSIS 2005, pp. 331–341. INSTICC Press, 2005

24. European Technical Standards Institute, `http://www.etsi.org/`

25. European Technical Standards Institute. GSM 11:11 - Digital cellular telecommunications system (phase 2+); Specification of the Subscriber Identity Module - Mobil Equipment (SIM-ME) interface, Version 8.3.0, 1999

26. European Technical Standards Institute, Security Algorithms Group of Experts (SAGE). `http://portal.etsi.org/sage/sage_tor.asp`

27. Europay International. MAOS Paltforms Status Technical Report, `www.europay.com`

28. K. Gandolfi, C. Mourtel, and F. Olivier. Electromagnetic analysis: concrete results. In Ç. K. Koç, D. Naccache and C. Paar, editors, *Cryptographic Hardware and Embedded Systems — CHES 2001*, volume 2162 of *Lecture Notes in Computer Science*, pp. 251–261. Springer-Verlag, 2001

29. Gemplus. MPCOS Multi Application Payment Chip, Reference Manual Ver 4.0, 1994

30. C. Giraud and H. Thiebeauld. A survey on fault attacks. In Y. Deswarte and A. A. El Kalam, editors, *Smart Card Research and Advanced Applications VI — 18th IFIP World Computer Congress*, pp. 159–176. Kluwer Academic, 2004

31. Global Platfom. Global Platform Card Specification, Version 2.1, 2001, `http://www.globalplatform.org`

32. L. Hemme. A differential fault attack against early rounds of (triple-)DES. In M. Joye and J.-J. Quisquater, editors, *Cryptographic Hardware and Embedded Systems — CHES 2004*, volume 3156 of *Lecture Notes in Computer Science*, pp. 254–267. Springer-Verlag, 2004

33. F. Hilebrand. GSM & UMTS, Wiley 2002

34. Intercede Group plc. OpenPlatform, http://www.intercede.com/Technology-OpenPlatform.htm

35. International Standard Organisation. ISO/IEC 7816, Information technology — Identification cards — Integrated circuit(s) cards with contacts — Part 4: Interindustry commands for interchange, 1995

36. International Standard Organisation. ISO/IEC 7816, Information technology — Identification cards — Integrated circuit(s) cards with contacts — Part 5: Numbering system and registration procedure for application identifiers, 1994

37. International Standard Organisation. ISO/IEC 7816, Information technology — Identification cards — Integrated circuit(s) cards with contacts — Part 6: Inter-industry data elements, 1996.

38. D.H Habing. The use of lasers to simulate radiation-induced transients in semi-conductor devices and circuits, In *IEEE Transactions On Nuclear Science*, volume 39, pp. 1647–1653, IEEE, 1992

39. Hive-Minded. Smartcard.NET, www.hiveminded.com

40. M. Joye and F. Olivier. Side-channel attacks. In H. van Tilborg, editor, *Encyclopedia of Cryptography and Security*, pp. 571–576. Kluwer Academic Publishers, 2005

41. M. Joye, J.-J. Quisquater, F. Bao, and R.H. Deng. RSA-type signatures in the presence of transient faults, In M. Darnell, editor, *Cryptography and Coding*, volume 1355 of *Lecture Notes in Computer Science*, pp. 155–160, Springer-Verlag, 1997

42. P. Kocher. Timing attacks on implementations of diffe-hellman, RSA, DSS, and other systems. In N. Koblitz, editor, *Advances in Cryptology — CRYPTO '96*, volume 1109 of *Lecture Notes in Computer Science*, pp. 104–113. Springer-Verlag, 1996

43. P. Kocher, J. Jaffe, and B. Jun. Differential power analysis. In M. J. Wiener, editor, *Advances in Cryptology — CRYPTO '99*, volume 1666 of *Lecture Notes in Computer Science*, pp. 388–397. Springer-Verlag, 1999

44. ITSEC. http://www.ssi.gouv.fr/site_documents/ITSEC/ITSEC-fr.pdf

45. T. M. Jurgensen and S. B. Guthery. Smart cards : the developer's toolkit, Prentice Hall, 2002

46. X. Leroy. Bytecode verification for java smart card. Software Practice & Experience, volume 32, pp. 319–340, 2002

47. MAOSCO Ltd. The MULTOS^{TM}Specification, http://www.multos.com/

48. C. Markantonakis. The case for a secure multi-application smart card operating system. In E. Okamoto, G. I. Davida, and M. Mambo, editors, *Information Security Workshop 97 — ISW '97)*, volume 1396 of *Lecture Notes in Computer Science*, pp. 188–197. Springer-Verlag, 1997

49. G. McGraw and E. W. Felten. Securing java, J. Wiley & Sons, 1999

50. G. McGraw, K. Ayer, and E. W. Felten. Jave Security meets smart cards, security enhancements in java card 2.1.1 will help multi-application smart cards take off in U.S. markets, Information Security Magazin, http://www.infosecurity.com/articles/march01/cover.shtml, 2001

51. T. S. Messerges. *Power Analysis Attacks and Countermeasures for Cryptographic Algorithms*. PhD thesis, University of Illinois, Chicago, 2000
52. T. S. Messerges. Using second-order power analysis to attack DPA resistant software. In Ç. K. Koç and C. Paar, editors, *Cryptogaphic Hardware and Embedded Systems — CHES 2000*, volume 1965 of *Lecture Notes in Computer Science*, pp. 71–77. Springer-Verlag, 2000
53. U. Meyer and S. Wetzel, On the Impact of GSM Encryption & Man-in-the-Middle Attacks on the Security of Interoperating GSM/YMTS Networks. In Proceedings of IEEE International Symposium on Personal, Indoor and Mobile Radio Communications — PIMRC 2004, volume 4, pp. 2876–2883, IEEE, 2004.
54. M. Montgomery, K. Krishna. Secure object sharing in Java card, In proceedings of the USENIX Workshop on Smart Card Tehnology — Smartcard '99, USENIX, 1999
55. D. Naccache, P. Q. Nguyẽn, M. Tunstall, and C. Whelan. Experimenting with faults, lattices and the DSA. In S. Vaudenay, editor, *Public Key Cryptography — PKC 2005*, volume 3386 of *Lecture Notes in Computer Science*, pp. 16–28. Springer-Verlag, 2005
56. General Information Systems Ltd. OSCAR, Specification of a smart card filling system incorporating data security and message authentication, `http://www.gis.co.uk/oscman1.htm`
57. Parliamentary Office of Science and Technology. Mobile Telephone Crime. In POST Briefing Note 64, 1995
58. J. R. Rao, P. Rohatgi, H. Scherzer, and S. Tinguely. Partitioning attacks: or how to rapidly clone some GSM cards. In Proceedings of IEEE Symposium on Security and Privacy, pp. 31–41, IEEE, 2002
59. E. Rose and K. H. Rose. Lightweight bytecode verification. In Formal Underpinnings of Java — OOPSLA '98, ACM, 1998
60. D. Samyde, S. P. Skorobogatov, R. J. Anderson, and J.-J. Quisquater. On a new way to read data from memory. In *Proceedings of the First International IEEE Security in Storage Workshop*, pp. 65–69, IEEE, 2002
61. D. Sauveron. *Étude et réalisation d'un environnemet d'expérimentation et de modélisation pour la technologie java card^{TM}. application à la sécurité*. PhD thesis, University of Bordeaux, Bordeaux, 2004
62. Season 2 Interface. `http://www.maxking.co.uk/`
63. S. P. Skorobogatov and R. J. Anderson. Optical fault induction attacks. In B. S. Kaliski Jr. and Ç. K. Koç and C. Paar, editors, *Cryptogaphic Hardware and Embedded Systems — CHES 2002*, volume 2523 of *Lecture Notes in Computer Science*, pp. 2–12. Springer-Verlag, 2002
64. SmartCard Trends. .NET brings web services to smart cards, April/May Issue, 2004
65. Sun Microsystems. Java Card API Ver 1.0, `www.javasoft.com/javacard/`
66. Sun Microsystems. Java Card API Ver 2.0, `www.javasoft.com/javacard/`
67. Sun Microsystems. Java Card 2.2.1 Application Programming Interface, 2003
68. Sun Microsystems. Java Card 2.2.1 Runtime Environment (JCRE) Specificqtion, 2003
69. Sun Microsystems. Java Card 2.2.1 Virtual Machine Specification, 2003
70. Sun Microsystems. Java Card API 2.2.1 Reference Implementation, 2002, `http://www.javasoft.com/products/javacard/`
71. Sun Microsystems. JSR 177 Expert Group. Security and Trust Services API (SATSA) for J2ME V1.0, 2004

72. D. A. Watt and D. F. Brown. Programming Language Processors in java: compilers and interpreters, Prentice Hall, 2000
73. M. Witteman, Java Card Security, Information Security Bulletin 8, pp. 291–298, 2003
74. ZeitControl. BasicCard. http://www.basiccard.com/
75. J. Ziegler. Effect of Cosmic Rays on Computer Memories, *Science*, volume 206, pp. 776–788, 1979

9

Governance of Information Security: New Paradigm of Security Management

Sangkyun Kim

SOMANSA, Seoul, Korea saviourkim@somansa.com; saviour@yonsei.ac.kr

Governance refers to the process whereby elements in society wield power and authority, and influence and enact policies and decisions concerning public life, and economic and social development [13]. There are three kinds of governance concept which should be considered in corporate environments: enterprise governance, IT governance, and security governance. The success factors of the governance are summarized: Adequate participation by business management; Clearly defined governance processes; Clarify stakeholders' roles; Measure the effectiveness of governance; Facilitate the evolution of governance; Clearly articulated goals; Resolution of cultural issues.

The approaches of security management, which manage an organization's security policy by monitoring and controlling security services and mechanisms, distribute security information, and report security events, are related with the purpose of security governance. However, studies on enterprise governance or IT governance, and security management lack in the provision of detailed framework and functionalities when considering the success factors of the governance described above. For example, BS7799, which is one of the most famous standards of security management in the world, provides general guidance on the wide variety of information security. Nevertheless, it takes the broad-brush approach. Accordingly, BS7799 does not provide definitive or specific materials on any topic of the security management and certainly could be useful as a high-level overview of information security topics that could help senior management to understand the basic issues involved in each of the topic areas.

This chapter provides a structured approach of security governance to corporate executives. Previous studies on the governance and security management are summarized to explain the components and requirements of a governance framework for corporate security. Finally, a governance framework for corporate security, which consists of four domains and two relationship categories, is provided. The domains have several objects respectively. The objects

S. Kim: *Governance of Information Security: New Paradigm of Security Management*, Studies in Computational Intelligence (SCI) **57**, 235–254 (2007)
www.springerlink.com

consist of components that should be resolved or provided to govern the issues of corporate security. The domains include a community (shareholder and management; media and customer; employee and supplier; government), security (control; enterprise strategy), performance (resource; competitive value), and information (owner; value; risk). The relationship among the objects of the security governance framework has two categories of harmonization and flywheel. The harmonization category governs the relationship among a community, performance, and security domain. The harmonization category deals with the problems of social, organizational, and human factors of corporate security. The flywheel category governs the relationship between a performance domain and security domain. The flywheel category deals with the virtuous cycle of corporate security. With this framework, corporate executives could create greater productivity gains, cost efficiencies, and a safer business community internally, for their customers and others interconnected throughout the critical infrastructure.

9.1 Introduction

Data security became computer security, computer security became IT security, IT security became information security and information security became business security because of the better understanding of the business impact and associated risk of not properly protecting a company's electronic resources [55]. However, it's difficult to resolve the problems of business security with the traditional approaches of corporate security management because traditional approach of corporate security management dose not focus on the governance of business impacts but concentrates on the value of information assets and risk analysis [32, 33, 36, 51, 60].

The need for governance exists anytime a group of people come together to accomplish an end. Most agree that the central component of governance is decision-making. It is the process through which this group of people make decisions that direct their collective efforts [01-20]. Recent studies of Conner and Coviello, IT Governance Institute, Moulton and Coles, Solms and Solms, Solms, Solms, Swindle and Conner show that the problems of business security should be resolved with the concept of security governance [8, 24, 41, 55–58].

This chapter provides an integrated framework of security governance. In the following sections, the literature on governance and security management are reviewed to explain fundamental perspectives of governance concept and business security. Second, the success factors of governance of corporate security are described according to the background and implications of governance concept. Third, the components of a governance framework for corporate security are explained along with the common approaches of governance frameworks and the limitations of security management. Finally, a governance framework for corporate security is provided with the domain, object, methods and practices of security governance.

9.2 Rise of the Governance

9.2.1 Definitions of the Governance

Governance refers to the process whereby elements in society wield power and authority, and influence and enact policies and decisions concerning public life, and economic and social development [13]. Yahoo's dictionary defines governance as "the act, process, or power of governing." There are three kinds of governance: enterprise governance, IT governance, and security governance. Related research and definition are summarized as described below.

Category: Enterprise Governance

Research institution or researcher: IT Governance Institute

Definition: Set of responsibilities and practices exercised by the board and executive management with the goal of providing strategic direction, ensuring that objectives are achieved, ascertaining that risks are managed appropriately and verifying that the enterprise's resources are used responsibly [23].

Research institution or researcher: OECD

Definition: It involves a set of relationships between a company's management, its board, its shareholders, and other stakeholders. It also provides the structure through which the objectives of the company are set, and the means of attaining those objectives and monitoring performance are determined. It should provide proper incentives for the board and management to pursue objectives that are in the interests of the company and its shareholders and should facilitate effective monitoring [45].

Category: IT Governance

Research institution or researcher: Allen

Definition: The actions required to align IT with enterprise objectives and ensure that IT investment decisions and performance measures demonstrate the value of IT toward meeting these [1].

Research institution or researcher: Appel

Definition: The principles, processes, people, and performance metrics enacted to enable freedom of thinking and action without compromising the overall objectives of the organization [3].

Research institution or researcher: IT Governance Institute

Definition: The leadership, organizational structures, and processes that ensure that the enterprise's IT sustains and extends the enterprise's strategies and objectives [23].

Research institution or researcher: Neela and Mahoney

Definition: IS(Information System) governance is the mechanism for assigning decision rights and creating an accountability framework that drives desirable behavior in the use of IT [42].

Category: Security Governance

Research institution or researcher: Moulton and Coles

Definition: It is the establishment and maintenance of the control environment to manage the risks relating to the confidentiality, integrity and availability of information and its supporting processes and systems [41].

9.2.2 Implications of the Governance

The benefits gained from introducing effective governance strategy to an IT department are considerable. The holistic view of IT governance integrates principles, people, processes, and performance into a seamless operational whole [3]. IT governance specifies the decision-making authority and accountability to encourage desirable behaviors in the use of IT. IT governance provides a framework in which the decisions made about IT issues are aligned with the overall business strategy and culture of the enterprise [9].

Solms stated that 'Information security is a direct corporate governance responsibility and lies squarely on the shoulders of the Board of the company [56].' Conner and Coviello stated that 'Corporate governance consists of the set of policies and internal controls by which organizations, irrespective of size or form, are directed and managed [8]. Information security governance is a subset of organizations' overall (corporate) governance program.' According to Solms, in the process of establishing environments for information security governance, companies are realizing that is it preferable to follow some type of internationally recognized reference framework for establishing an information security governance environment, rather than doing it ad hoc [57]. By using the information security governance framework, CEOs and boards of directors will create a safer business community internally and for their customers and others interconnected throughout the critical infrastructure. In aggregate, such measures serve as an executive call to action that will also help better protect our nation's security. Information security governance will provide organizations far greater benefits than just legal or regulatory compliance. Robust security serves as a catalyst to even greater productivity gains and cost efficiencies for businesses, customers, citizens and governments during times of crisis and normal operations [58].

IT Governance Institute summarized the necessities of security governance: Risks and threats are real and could have significant impact on the enterprise; Effective information security requires coordinated and integrated action from the top down; IT investments can be very substantial and easily misdirected; Cultural and organizational factors are equally important; Rules and priorities need to be established and enforced; Trust needs to be demonstrated toward trading partners while exchanging electronic transactions; Trust in reliability in system security needs to be demonstrated to all stakeholders; Security incidents are likely to be exposed to the public; Reputational damage can be considerable [24].

9.2.3 Success Factors of the Governance

Dallas proposed six steps which should be took by CIOs to improve IT governance effectiveness. Dallas's model includes: 1. Match the governance initiative with the decision-making style of the enterprise; 2. Align decision-making authority with the domain; 3. Integrate the governance mechanisms - that is, the structures and processes - and evolve them; 4. Clarify stakeholders' roles; 5. Measure the effectiveness of IT governance; 6. Facilitate the evolution of IT governance [10].

Avoid the common causes of IT governance failure by collaborating with business leaders to create processes designed to achieve governance goals tailored to your enterprise's culture and management style. This failure to implement usually results from one or more of the following three factors: inadequate participation by business management; a lack of clearly articulated goals; a lack of clearly defined governance processes [15].

An organization's culture is the common values and beliefs through which it views its people, customers, opportunities and decisions. Organizational culture is a powerful influence on the success or failure of governance mechanisms. Cultures that allow bad governance usually rely more on power than process [42].

To sum up, the success factors of the governance are summarized as described below:

No.1 Adequate participation by business management (align decision-making authority with the domain)

No.2 Clearly defined governance processes (integrate the governance mechanisms and evolve them)

No.3 Clarify stakeholders' roles

No.4 Measure the effectiveness of governance

No.5 Facilitate the evolution of governance

No.6 Clearly articulated goals

No.7 Resolution of cultural issues

9.3 Why the Security Management Fails

9.3.1 What the Security Management Can Do

Bitpipe defines security management as "the management of an organization's security policy by monitoring and controlling security services and mechanisms, distributing security information, and reporting security events." Table 9.1 summarizes the functions and descriptions of the security management.

Among the studies on the security management, BS7799 is the most widely recognized security standard in the world which includes the whole realm of security management. BS7799 is the most widely recognized security standard in the world. Although it was originally published in the mid-nineties, it was revised in May 1999 which really put it on to the world stage. Ultimately, it evolved into BS EN ISO17799 in December 2000. BS7799(ISO17799) is comprehensive in its coverage of security issues containing a significant number of control requirements. Compliance with BS7799 is consequently far from trivial task, even for the most security consciousness of organizations. It consists of ten major sections, each covering a different topic or area.

Business Continuity Planning: The objectives of this section are summarized: To counteract interruptions to business activities and to critical business processes from the effects of major failures or disasters.

Access Control: The objectives of this section are summarized: 1) To control access to information, 2) To prevent unauthorized access to information systems, 3) To ensure the protection of networked services 4) To prevent unauthorized computer access, 5) To detect unauthorized activities, and 6) To ensure information security when using mobile computing and tele-networking.

System Development and Maintenance: The objectives of this section are summarized: 1) To ensure security is built into operational systems, 2) To prevent loss, modification or misuse of user data in application systems, 3) To protect the confidentiality, authenticity and integrity of information, 4) To ensure IT projects and support activities are conducted in a secure manner, and 5) To maintain the security of application system software and data.

Physical and Environmental Security: The objectives of this section are summarized: 1) To prevent unauthorized access, damage and interference to business premises and information, 2) To prevent loss, damage or compromise of assets and interruption to business activities, and 3) To prevent compromise or theft of information and information processing facilities.

Table 9.1. What the security management can do

Function	Related research	Description
Auditing of information security management system	BS7799; ISO17799 [22]	10 domains of evaluation and auditing factors of information security management system
Evaluation of information security management system	Kim and Leem [28]; NIST [44]; Tudor [59]	Evaluation factors of information security management system
Life cycle of security activities	Henze [18]; ISO13335 [21]; Kim et al. [34]; NIST [43]	Process model of enterprise security management
Security strategy planning	Kim and Leem [29]; Kim and Leem [32]	Methodology for strategy planning of enterprise security systems
Domain and capability of security engineering	SEI [54]	Maturity model of information security management system
Risk analysis	Kim and Leem [31]; Kim and Leem [33]; Rex, Charles and Houston [49]; Ron [50]	Analysis and assessment of risks including asset value, vulnerability, and threat
Evaluation of security products	Gilbert [01-16]; Kim and Leem [30]; Lynch and Stenmark [37]; Polk and Bassham [46]	Evaluation and selection process of security controls
Classification, decision and comparison factors of security controls	TCSEC; ITSEC; CC; Beall and Hodges [5]; Firth, et al. [11]; Fites et al. [12]; Hutt [19]; Kavanaugh [26]; Kim and Leem [30]; Kim [35]; Krutz and Vines [36]; Schweitzer [52]; Vallabhaneni [60]	Evaluation and selection factors of security controls
Concept of economic justification	Blakley [6]; Geer [14]; Malik [38]; Scott [53]	Framework of economic justification of security investment
Empirical study on economic justification	Bates [4]; Power [48]	Process of economic justification of security investment
Cost factors of security systems	Harris [17]; Roper [51]; Witty [62]	Cost factors of economic justification of security investment

Compliance: The objectives of this section are summarized: 1) To avoid breaches of any criminal or civil law, statutory, regulatory or contractual obligations and of any security requirements, 2) To ensure compliance of systems with organizational security policies and standards, and 3) To maximize the effectiveness of and to minimize interference to/from the system audit process.

Personnel Security: The objectives of this section are summarized: 1) To reduce risks of human error, theft, fraud or misuse of facilities, 2) To ensure that users are aware of information security threats and concerns, and are equipped to support the corporate security policy in the course of their normal work, and 3) To minimize the damage from security incidents and malfunctions and learn from such incidents.

Security Organization: The objectives of this section are summarized: 1) To manage information security within the Company, 2) To maintain the security of organizational information processing facilities and information assets accessed by third parties, and 3) To maintain the security of information when the responsibility for information processing has been out-sourced to another organization.

Computer and Network Management: The objectives of this section are summarized: 1) To ensure the correct and secure operation of information processing facilities, 2) To minimize the risk of systems failures, 3) To protect the integrity of software and information, 4) To maintain the integrity and availability of information processing and communication, 5) To ensure the safeguarding of information in networks and the protection of the supporting infrastructure, 6) To prevent damage to assets and interruptions to business activities, and 7) To prevent loss, modification or misuse of information exchanged between organizations.

Asset Classification and Control: The objectives of this section are summarized: To maintain appropriate protection of corporate assets and to ensure that information assets receive an appropriate level of protection.

Security Policy: The objectives of this section are summarized: To provide management direction and support for information security.

BS7799 provides general guidance on the wide variety of topics listed above. It takes the broad-brush approach. So BS7799 does not provide definitive or specific materials on any topic of the security management and certainly could be useful as a high-level overview of information security topics that could help senior management to understand the basic issues involved in each of the topic areas.

9.3.2 What the Security Management Cannot Do

By 'elevating' Information Security to Business Security, it will get the extra focus and attention it needs to ensure the prolonged existence of the company, and to integrate all the present efforts as far as such protection is concerned [55]. However, the functions of the security management have some limitations to solve the problems of business security as described in the sequel.

Success factors of governance no.1

Description: Adequate participation by business management (align decision-making authority with the domain)

Related research based on the security management: Kim and Leem [29]; Kim and Leem [30]; Kim and Leem [32]

Limitations (Additional requirements of the security governance): Focuses on the provision of decision-making factors; Concentrates on the technology-oriented factors.

Success factors of governance no.2

Description: Clearly defined governance processes (integrate the governance mechanisms and evolve them)

Related research based on the security management: ISO13335; Henze [18]; Kim et al. [34]; NIST [43]

Limitations (Additional requirements of the security governance): Focuses on the provision of lifecycle of security controls; Lacks in the integration of the governance mechanism and various participants of enterprise security.

Success factors of governance no.3

Description: Clarify stakeholders' roles

Related research based on the security management: Not available

Limitations (Additional requirements of the security governance): Does not provide the relationship and responsibility of stakeholders.

Success factors of governance no.4

Description: Measure the effectiveness of governance

Related research based on the security management: Bates [4]; Blakley [6]; Geer [14]; Harris [17]; Malik [38]; Power [48]; Roper [51]; Scott [53]; Witty [62]

Limitations (Additional requirements of the security governance): Lacks in the provision of strategic benefits and overall performance.

Success factors of governance no.5

Description: Facilitate the evolution of governance

Related research based on the security management: SEI [54]

Limitations (Additional requirements of the security governance): Only provides the maturity model of security engineering (doest not provide a virtuous cycle of enterprise security).

Success factors of governance no.6

Description: Clearly articulated goals

Related research based on the security management: Kim and Leem [29]; Kim and Leem [31]; Kim and Leem [32]; Kim and Leem [33]; Rex, Charles and Houston [49]; Ron [50]

Limitations (Additional requirements of the security governance): Lacks in the alignment of security strategy with business strategy.

Success factors of governance no.7

Description: Resolution of cultural issues

Related research based on the security management: Not available

Limitations (Additional requirements of the security governance): Does not consider the relationship among the participants of enterprise security.

9.4 Governance of Corporate Security

9.4.1 General Frameworks for the Governance

There are various frameworks on the governance. Table 9.2 summarizes the existing studies on a governance framework including the enterprise governance, IT governance, and security governance.

There are also three frameworks which show a fundamental concept of security governance as described in table 9.2. However, these frameworks on the enterprise governance, IT governance, and security governance do not sufficiently meet the requirements of the success factors of security governance. Table 9.3 summarizes the limitations of general frameworks which lack in the provision of an essential architecture, domain, and presentation for the governance of corporate security.

9.4.2 Integrated Framework for the Governance of Corporate Security

Table 9.4 summarizes the key factors of a governance framework for corporate security considering the success factors of the governance, what the security management cannot do, and the limitations of general frameworks.

Table 9.2. General frameworks for the governance

Category	Research	Component
Enterprise governance	Connell et al. [7]	Conformance processes; Performance processes; Accountability and Assurance; Value creation and Resource utilization
IT governance	Weill and Woodham [61]	Component: Business objectives (desirable behavior), IT governance style (IT governance mechanisms), Business performance goals (metrics), IT domains (IT principles, IT infrastructure, IT architecture, IT investment); Relationship: harmonize what, harmonize how
	CobiT [40]	IT processes; IT information criteria; IT resources
	Mercury Interactive Corporation	Demand management; Portfolio management; Program management; Software demand; Operational demand; Deployment; Change management
Security governance	Moulton and Coles [41]	Security responsibilities and practices; Strategies/objectives for security; Risk assessment and management; Resource management for security; Compliance with legislation, regulations, security policies and rules; Investor relations and communications activity in relation to security
	IT Governance Institute [24, 25]	Strategic Alignment; Value Delivery; Risk Management; Performance Measurement
	Posthumus and Solms [47]	Component: Business issues, Legal/regulatory, IT infrastructure, Standards/best practices, Management (vision, strategy, mission, policy); Relation: direct, control, information risk directive

Table 9.3. Limitations of general frameworks

Previous framework	Limitations
Connell et al. [7]	Does not provide the detailed process and community model of enterprise security
Weill and Woodham [61]	Security issues are treated as a partial element of IT management
CobiT [40]	Fundamentally, its a IT auditing methodology; Lacks in the alignment with security strategy with enterprise strategy
Mercury Interactive Corporation	Lacks in the consideration of business security; Focuses on the lifecycle management of enterprise information systems
Moulton and Coles [41]	Performance management and detailed description of security controls are required
IT Governance Institute [24]	Management of the relationships among the participants of enterprise security is required; Factors of investment and performance should be described in further detail
Posthumus and Solms [47]	Relationship among the participants of enterprise security, detailed descriptions of governance components, investment and performance management are required

Table 9.4. Key factors of the governance framework for corporate security

View	Key factors
Architecture	Domain which adjusts the partitions of every object of enterprise security; Clear relationship among domains
Domain	Every participants of enterprise security should be considered; Characteristics of business information; Performance management should provide the cost and benefit factors; Security controls and strategies should be subdivided in further detail
Presentation	Illustration which shows a bird-eye view of the security governance framework should be provided; Structured table which describes every object and component of domain, and their characteristics should be provided

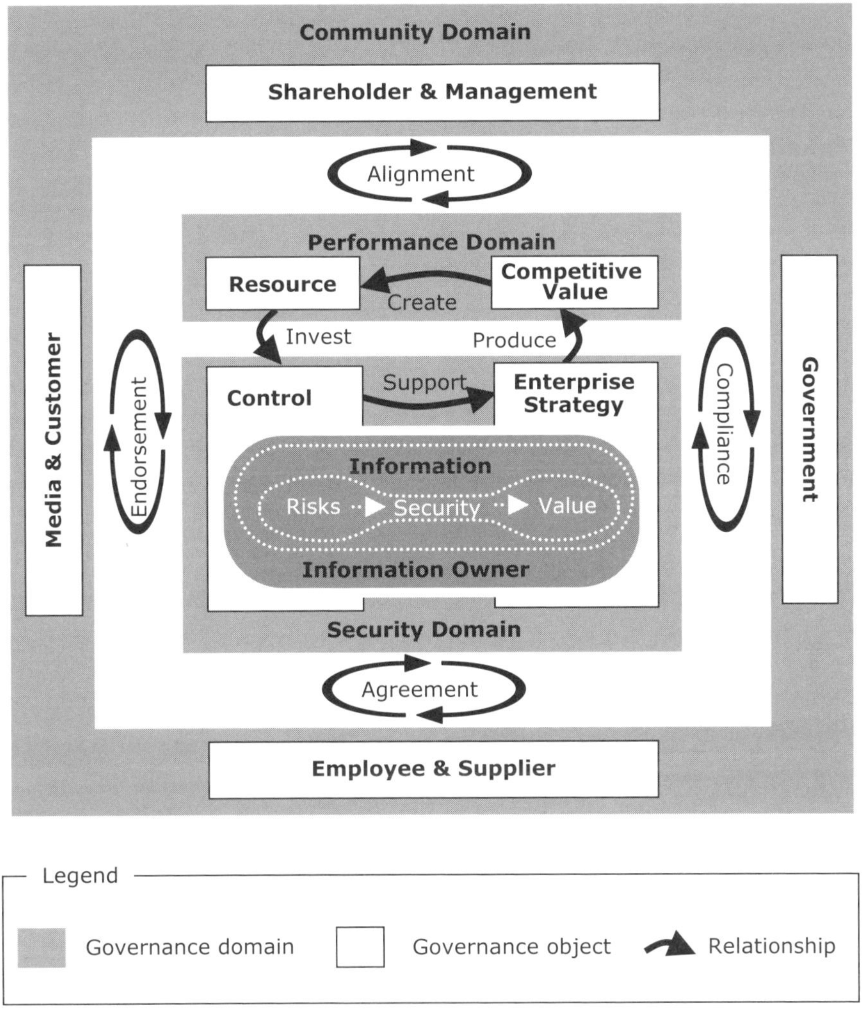

Fig. 9.1. Security governance framework

Considering the key factors of table 9.4, a framework of the security governance is provided as illustrated in figure 9.1.

The governance framework for corporate security consists of four domains and two relationship categories. The domains include: Community, Security, Performance, and Information. With the domains, every participants of enterprise could be considered in governing information security. The domains are correlated with a fundamental consideration of performance management including the cost and benefit factors. The domains have several objects respectively. The objects consist of some components that should be resolved or provided to govern the corporate security. The characteristics of the domains, objects, and components of the security governance framework are summarized as described below.

Domain - Community

Object - Shareholder and Management

Component - Vision: The ultimate mission of the enterprise

Component - Philosophy: The philosophical principles of management and investors

Object - Media and Customer

Component - Emotion: Mental response, which arises spontaneously rather than through conscious effort, of customers on the external images of enterprise's security governance

Component - Public opinion: Public consensus, as with respect to an principles of enterprise's security governance

Object - Employee and Supplier

Component - Culture: The common values and beliefs, and predominating attitudes and behavior of employees

Component - Awareness: Having knowledge or cognizance on the strategy and principles of their enterprise security [60]

Object - Government

Component - Legislation: Criminal or civil law, statutory, regulatory or contractual obligations related with enterprise security including protection of asset, privacy, and so on (ISO17799)

Component - Standard: Set of agreements produced by many organizations; BS7799/ISO17799, ISO13335, ISO7816, ISO9798, ISO10736, ISO11577, ISO14765, ISO9594, ISO10164, ISO10181, ISO8372, ISO9796, ISO10118, and ISO13888 are widely recognized standards of information security

Domain - Security

Object - Control

Component - Policy: General statement produced by senior management to dictate what type of role security plays within the organization [17]

Component - Standard: Specification of the use of specific technologies in a uniform way [36]

Component - Procedures: The detailed steps that are followed to perform a specific task; The detailed actions that personnel must follow [36]

Component - Guideline: Discretionary statement or other indication of security policy or procedure by which to determine a course of action

Component - Responsibilities: A duty, obligation, or burden that each member of the enterprise should keep to achieve the goal and objectives of enterprise's security strategy

Component - System: Technical controls including an access control, encryption, backup, auditing, and intrusion detection

Component - Facility: HVAC and physical access control

Object - Enterprise Strategy

Component - Business strategy: It's the basis for the organization's existence and includes a mission statement which defines the business goals and objectives that typically include references to revenue growth, increased production and performance, competitive advantage, increased quality and service levels, and decreased costs and time-to-market [59]

Component - Information strategy: It's the part of business strategy concerned with deploying the enterprise's information systems resources, including people, hardware, and software [2]

Component - Security strategy: It defines the mission, goals, and objectives of enterprise security management

Domain - Performance

Object - Resource

Component - Facility: Building, geographic location, environmental control and physical safety [36]

Component - Organization: A group of persons organized for security governance; A structure through which individuals cooperate systematically to govern the enterprise security

Component - System: Network, server, and applications

Component - Data: Business data, information, knowledge, and intelligence

Component - Time: Priority of tactical or operational planning

Component - Motivation: Provision of an incentive to improve the agreement and awareness of security principles

Object - Competitive Value [27]

Component - Operational benefits: Enhanced efficiency of organization's operations including cost saving, added profitability, enhanced decision-making and enhanced business function

Component - Strategic benefits: Enhanced competitive advantages including the reduced threat of rivalry, enhanced supplier relationship, and enhanced customer relationship

Domain - Information

Object - Owner

Component - Personnel: Member of the enterprise who creates, processes, memorizes, and communicates the business information

Component - Facility: Media; Environments for media or personnel and system

Component - System: Assembly of network, server, and applications which is used to create, process, store, and share the business information

Object - Value

Component - Official value: Market value of information asset including the cases of exclusive possession, liability, book value, and expert appraisal

Component - Utility: Potential value of information asset for the operational or strategic utilities

Object - Risk [39]

Component - Confidentiality: Protection from unauthorized disclosure

Component - Integrity: Protection from unauthorized, unanticipated, or unintentional modification

Component - Availability: Available on a timely basis to meet mission requirements or to avoid substantial losses.

The relationship among the domains of the security governance framework has two categories of harmonization and flywheel. The characteristics of each relationship are summarized as described below.

Category of Relationship - Harmonization

Relationship - Alignment: The ultimate goal of business security is not a safeguard for the enterprise's information but an improvement of competitive value from the effective support for the enterprise's strategies [35]

Relationship - Agreement: In spite of the importance of support and conformation of employees and suppliers, they seldom have an opportunity to design and implement the enterprise security strategy; Mutual agreement between the management body of enterprise security and employees and suppliers should be established [17]

Relationship - Endorsement: Principles of security governance should not go against the stream of popular opinion [35]

Relationship - Compliance: To avoid breaches of any criminal or civil law, statutory, regulatory or contractual obligations, it should be checked continuously (BS7799; ISO17799)

Category of Relationship - Flywheel

Relationship - Invest: Abundant resources delivered from the enhanced competitiveness are invested to plan, implement, and operate the security controls

Relationship - Support: Security controls achieve the goal of security strategy which is a subset of the business and information strategy [29,32]

Relationship - Produce: Achievement of the goal of security strategy produces an enhanced competitiveness including operational and strategic benefits [27]

Relationship - Create: Enhanced competitiveness resulted from the security governance delivers more abundant resources for security governance.

The harmonization category governs the relationship among the performance, security, and community domain. The harmonization category deals with the problems of social, organizational, and human factors of corporate security. The flywheel category governs the relationship between the performance domain and security domain. The flywheel category deals with the virtuous cycle of corporate security.

9.5 Summary

This chapter provides a governance framework of corporate security. Following studies are summarized: 1) Definition of the enterprise governance, IT governance, and security governance. 2) Literatures on the security management. 3) General frameworks for the governance. Finally, a framework for the security governance, which consists of four domains including eleven objects and two relative categories including eight relationships, is provided.

Managerial implications of the security governance framework are summarized as follow: 1) It integrates strategies, security controls, performance, and communities into a seamless operational whole. 2) It specifies the decision-making authority and accountability to encourage desirable behaviors. 3) Robust governance of enterprise security creates greater productivity gains and cost efficiencies for businesses.

References

1. Allen J (2005) An introduction to governing for enterprise security. Software Engineering Institute, Carnegie Mellon University in Pittsburgh
2. Alter S (1999) Information systems: a management perspective. Addison-Wesley, New York
3. Appel W (2005) Redefining IT governance readiness. META Group
4. Bates RJ (1991) Disaster recovery planning. McGraw-Hill, New York
5. Beall S, Hodges R (2002) Protection and security: software comparison columns. Gartner Inc., Stamford
6. Blakley B (2001) Returns on security investment: an imprecise but necessary calculation. Secure Business Quarterly 1
7. Connell B, Rochet P, Chow E, Savino L, Payne P (2004) Enterprise governance: getting the balance right. International Federation of Accountants
8. Conner FW, Coviello AW (2004) Information security governance: a call to action. National Cyber Security Summit Task Force
9. Dallas S, Bell M (2004) The need for IT governance: now more than ever. Gartner Inc., Stamford
10. Dallas S (2002) Six IT governance rules to boost IT and user credibility. Gartner Inc., Stamford
11. Firth R, Fraser B, Konda S, Simmel D (1998) An approach for selecting and specifying tools for information survivability. Software Engineering Institute, Carnegie Mellon University in Pittsburgh
12. Fites PE, Kratz MPJ, Brebner AF (1989) Controls and security of computer information systems. Computer Science Press, Rockville
13. GDRC (2005) The global development research center (http://www.gdrc.org)
14. Geer DE (2001) Making choices to show ROI. Secure Business Quarterly 1
15. Gerrard M (2003) Creating an effective IT governance process. Gartner Inc., Stamford
16. Gilbert IE (1989) Guide for selecting automated risk analysis tools (SP 500-174). NIST, Gaithersburg
17. Harris S (2003) CISSP all-in-one exam guide 2nd edition. McGraw-Hill, New York
18. Henze D (2000) IT baseline protection manual. Federal Agency for Security in Information Technology, Germany
19. Hutt AE (1988) Management's roles in computer security. In: Hutt AE (eds) Computer security handbook. Macmillan, New York
20. Institute on Governance (2005) What is governance?: getting to a definition (http://www.iog.ca)
21. ISO13335-1: information technology - guidelines for the management of IT security - part 1: concepts and models for IT security. International Organization for Standardization, Geneva
22. ISO17799: information technology - security techniques - code of practice for information security management. International Organization for Standardization, Geneva
23. IT Governance Institute (2001) Board briefing on IT governance. IT Governance Institute, Rolling Meadows
24. IT Governance Institute (2004) Information security governance. IT Governance Institute, Rolling Meadows

25. IT Governance Institute (2001) Information security governance: guidance for boards of directors and executive management. IT Governance Institute, Rolling Meadows
26. Kavanaugh K (2001) Security services: focusing on user needs. Gartner Inc., Stamford
27. Kim S, Lee HJ (2005) Cost-benefit analysis of security investments: a methodology and case study. Lecture Notes in Computer Science 3482: 1239-1248
28. Kim S, Leem CS (2004) An evaluation methodology of enterprise security management systems. Fifth International Conference on Operations and Quantitative Management, Seoul
29. Kim S, Leem CS (2004) An information engineering methodology for the security strategy planning. Lecture Notes in Computer Science 3482: 597-607
30. Kim S, Leem CS (2004) Decision supporting method with the analytic hierarchy process model for the systematic selection of COTS-based security control. Lecture Series on Computer Science and on Computational Sciences 1: 896-899
31. Kim S, Leem CS (2004) Implementation of the security system for instant messengers. Lecture Notes in Computer Science 3314: 739-744
32. Kim S, Leem CS (2004) Information strategy planning methodology for the security of information systems. ICCIE 2004, Cheju
33. Kim S, Leem CS (2005) Security of the internet-based instant messenger: risks and safeguards. Internet Research: Electronic Networking Applications and Policy 15: 88-98
34. Kim S, Choi SS, Leem CS (1999) An integrated framework for secure e-business models and their implementation. INFORMS'99, Seoul
35. Kim S (2002) Security consultant training handbook. HIT, Seoul
36. Krutz RL, Vines RD (2001) The CISSP prep guide: mastering the ten domains of computer security. John Wiley and Sons, New York
37. Lynch G., Stenmark I (1996) A methodology for rating security vendors. Gartner Inc., Stamford
38. Malik W (2001) A security funding strategy. Gartner Inc., Stamford
39. Marianne S (1998) Guide for developing security plans for information technology systems. NIST, Gaithersburg
40. Mingay S, Bittinger S (2002) Combine CobiT and ITIL for powerful IT governance. Gartner Inc., Stamford
41. Moulton R, Coles RS (2003) Applying information security governance. Computers and Security 22: 580-584
42. Neela AM, Mahoney J (2003) Work with, not against, your culture to refine IT governance. Gartner Inc., Stamford
43. NIST (1995) An introduction to computer security: the NIST handbook. NIST, Gaithersburg
44. NIST (2001) Security self-assessment guide for information technology systems. NIST, Gaithersburg
45. OECD (1999) OECD principles of corporate governance. Organization for Economic Cooperation and Development
46. Polk WT, Bassham LE (1992) A guide to the selection of anti-virus tools and techniques (SP 800-5). NIST, Gaithersburg
47. Posthumus S, Solms RV (2004) A framework for the governance of information security. Computers and Security 23: 638-646
48. Power R (2002) CSI/FBI computer crime and security survey. Federal Bureau of Investigation, Washington

49. Rex RK, Charles SA, Houston CH (1991) Risk analysis for information technology. Journal of Management Information Systems 8
50. Ron W (1988) EDP auditing: conceptual foundations and practice. McGraw-Hill, New York
51. Roper CA (1999) Risk management for security professionals. Butterworth-Heinemann, Boston
52. Schweitzer JA (1983) Protecting information in the electronic workplace: a guide for managers. Reston Publishing Company, Reston
53. Scott D (1998) Security investment justification and success factors. Gartner Inc., Stamford
54. SEI (1999) A systems engineering capability maturity model version 2. Software Engineering Institute, Carnegie Mellon University in Pittsburgh
55. Solms BV, Solms RV (2005) From information security to business security?. Computers and Security 24: 271-273
56. Solms BV (2001) Corporate governance and information security. Computers and Security 20: 215-218
57. Solms BV (2005) Information security governance: CobiT or ISO 17799 or both?. Computers and Security 24: 99-104
58. Swindle O, Conner B (2004) The link between information security and corporate governance. Computerworld
59. Tudor JK (2000) Information security architecture: an integrated approach to security in the organization. Auerbach, New York
60. Vallabhaneni R (2000) CISSP examination textbooks. SRV Professional Publications, Los Angeles
61. Weill P, Woodham R (2002) Don't just lead, govern: implementing effective IT governance. Center for Information Systems Research, Sloan School of Management, Massachusetts Institute of Technology in Cambridge
62. Witty RJ, Girard J, Graff JW, Hallawell A, Hildreth B, MacDonald N, Malik WJ, Pescatore J, Reynolds M, Russell K, Wheatman V, Dubiel JP, Weintraub A (2001) The price of information security. Gartner Inc., Stamford

Author Index

Printing: Krips bv, Meppel
Binding: Stürtz, Würzburg